SEA SEARCH

By the Editors of
Treasure Seekers Magazine

Sea Search

Sea Search

Copyright © 2004 by
Graphic Image Publications
P.O. Box 1395
Agoura Hills, California
91376-1395 U.S.A.
graphicimage454@aol.com

ISBN 0-9749971-0-2

All rights reserved. This book, or parts thereof, may not be reproduced,
stored in a retrieval system or transmitted in any form
without written permission from the publisher.

*** *CAUTION NOTICE* ***

State, federal and antiquities laws are subject to change. The published summaries are not to be considered as legal advice or a reinstatement of the law. To determine the applicability of these laws to specific situations which you may encounter, you are strongly urged to consult a local attorney. The publisher cannot verify the authenticity or validity of every story contained herein. Should any reader have a problem with the information offered, he/she should contact local authorities and receive the proper interpretation of local laws prior to the initiation of a search.

Sea Search

Sea Search

Contents

The *Lexington*, Long Island's First Sea Disaster 9
Escape From The *S-5* 17
Treasure Fleets of Spain:
    Where Did They Come From? 25
The *President Coolidge* 33
Morro Castle, Ship Of Doom! 39
Oak Island, The Home Of Pirate Treasure! 47
Sultana, A Ship That Was A Flaming Coffin 55
The Ghost Of *Bannockburn* 67
The Fate Of The *General Slocum* 71
The Elusive *City of Rio de Janeiro* 79
Disappearance Of The Ill-Fated *Portland* 87
Lady Liza, The Little Tramp 93
The Ship That Returned From The Dead........................101
The Treasure Ship *Brother Jonathan* 111
The Salvage Miracle At Pearl Harbor 121
The Lost Treasure Of Manila Bay 127
Final Voyage Of The *S.S. Pacific* 143
Treasure Ships Of New England 151
The Big Boom In Treasure Salvage 159
Hey Look! It's The *Great Eastern* 173
A Whale Of A Story..181
Christopher Columbus, Salesman!...................................187
The *Eastland,* Perished Where Berthed 195
The *Arabia,* She's A Steamboat A' Comin'! 201

Sea Search

Introduction

Welcome to *Sea Search*. The following historical recordings are a compendium of sea collectibles for landlubbers and professional seafarers.

As editor, I've had an interest in the sea with its lost treasures resting beneath its surface ever since I was a child living on Long Island. Growing older, I eventually became the captain of a small Corinthian sailboat — a sprawling 20 feet in length sailing the Great South Bay of Long Island!

While I was undenyabably aware that my single-handed nautical skills were slightly less than those of Magellan or Captain Cook, the thrill of being at sea provided the same exhilaration my boyhood heroes experienced.

It was inevitable — while sailing alone, thoughts came to mind of the thousands of sunken ships whose skeletal remains rest on the bottom of the sea. Most of these victims carried precious cargo that still lie within the holds.

My adrenaline rose! Suspecting there is more significance to these forgotten stories than just thoughts of the hidden untold wealth, I wanted to know more! The vessels' historical value, their missions and the facts that led to their demise all play an equal part of importance to the mystical tales surrounding their being. My fascination with these yarns still lingers today.

Recognizing others have experienced the same inquisitive stimulus, we decided to put this book together. *Sea Search* offers a broad view of doomed ships and sunken lost treasure. The countless hours of research, dives and underwater adventure that follow will not be found in other publications.

Our contributors focus on a broad scope of nautical interest: history, diving, treasure and the ships themselves.

We hope you get as much enjoyment out of reading this work as the writers did in compiling the stories that follow. Every effort was made to research the data supplied and to bring you interesting factual information.

Our thanks to those who provided the research, writing and editing of this book: Keith Bombard, Percy Thawnton, Paul McCallum, Alfred Pawling, Jeff Hollingsworth, Ed Schnepf, Cecil Rhodes, Robert K. Mann, Paul Bowers, Palmer A Jackson and Owen Gault.

Richard C. Schnepf, Editor

The *Lexington*, Long Island's First Sea Disaster

With the birth of the industrial age in the early nineteenth century, steam-engine concepts became a reliable source of power. It was Robert Fulton who proved steam power installed within a ship's hull was the ultimate replacement for the sail. Staunche shipbuilders, however, were not totally convinced of steam engineering designs. Ultimately they knew that ships' architecture had to change. Engineers of the day began incorporating steam-generating power with sail. Reliability was their first concern. The first tests proved that sails not only added assurance to a voyage, but were an aid to the overall ship's speed and performance. Sailing ships with side-wheel paddles slowly took their position on the seas as a safe means of transportation.

In 1835 Cornelius Vanderbilt placed an order with Simionson & Bishop, shipbuilders of Brooklyn, New York, for a newly designed 488-ton side-wheeler passenger/steamboat. The new ship, christened the *Lexington*, operated completely by steam and without the assistance of sails. The vessel's design measured 220 feet long with a narrow 26-foot beam. Her draft cut the water at 11 feet. The *Lexington's* compound steam engine and boiler, purchased from the West Point Foundry Company of New York, provided her with a remarkable 13-knots of speed.

Upon completion of the vessel and proud of his acquisition, Vanderbilt knew his company was the first to offer scheduled passenger/freight service to the New England manufacturing ports from New York. The purchase of a ticket guaranteed safe, reliable passenger service for 150-plus persons. The *Lexington's* speed plus its capability to carry large amounts of freight minimized the arduous horse-drawn trip from New York to the New England coast. The commodore was thrilled with first reports that the *Lexington's* maiden voyage between New York and Providence, Rhode Island, was completed in less than 12 hours.

The ship exceeded all expectations of the builder. Although the ship was designed as a dayliner/freighter, Vanderbilt insisted that comfortable accommodations be included in its design. Several small stateroom cabins with berths placed along the forward portion of the ship provided a comfortable, relaxed atmosphere. The four dollar ticket that sat passengers on hard wooden benches could be easily upgraded to soft beds during an overnight passage. Food and beverage service was also available at a modest additional cost. Rest rooms, although still used on a communal basis, were advertised as clean and airy. The ship, considered in 1835 to be modern and comfortable, became the pride of Vanderbilt's fleet. Captain George Child was the man selected by Vanderbilt to master his new ship.

Monday morning, January 13, 1840, Captain Child of Narragansette, Rhode Island, opened his eyes to a dismal overcast day that threatened snow and a frigid temperature of 10 degrees below zero. Trying to warm his chilled body, he boarded the *Lexington* to prepare for the ship's passage to Providence, Rhode Island. En route, he had a planned stop in Stonington, Connecticut, to unload passengers and cargo.

Captain Child was concerned about the penetrating low morning temperature. He feared freezing of the East River and Long Island Sound waters. The ship's wooden side-wheel paddles would surely be damaged if they had to plow through icy crusts. With this in mind, Child wanted to get away by mid-afternoon.

The *Lexington*

Longshoremen, having problems with frozen winches and tackle complained of the bitter cold. Palletized freight stuck to the dock as though it were welded with ice. Cargo movement required pry-bars to break the skids from frozen surfaces.

Slowly, the cotton bales brought from the South for delivery to New England textile mills were loaded into the ship's hold. Some were haphazardly placed and lashed to the deck. Despite his knowing the temperature hampered the workers, Captain Child was irritated by their slow actions and indifference to his commands. His patience was being tested.

At three in the afternoon, a light snow started to fall and the temperature hovered well below zero. Child was handed the cargo manifests and informed all passengers were aboard. The engine room crew awaited commands from the bridge. Child gave the order to drop dock lines and get underway. As he anticipated, the *Lexington* made slow progress through the ice crusted East River.

Although the actual number of passengers was never recorded, most referenced reports show 118 persons to be aboard. Several were returning to New England from their Christmas holiday in New York. Child was impressed by his passenger list. It included three of his peers, all seasoned seafarers: Ship Captains Bertran T. Foster, E.J. Kimball and Icabod D. Carver.

Carver's scarf-covered face gleamed with joy when he boarded — he was about to be married the next day. Prominent businessmen from New York and Boston were also on board. Alophus Harnden, superintendent of Harnden's Express Co., carefully packed $50,000 in bank notes, $20,000 in silver coin and another $70,000 in silver bars into a safe. The safe was securely chained to the ship's deck by two crewmen and double checked by the superintendent for added safety. The money was scheduled for delivery at Harnden's bank in Providence for deposit. Author/poet Henry Wadsworth Longfellow's name also appeared on the manifest. As fate would have it, he sold his poem *The Wreck of the Hesperus* that morning for $25. His new found notoriety changed his plans, cancelling his trip. Longfellow decided to give several lectures in New York instead of traveling to Rhode Island.

Passengers huddled around the ship's stoves trying to warm their chilled bodies. Knowing that the trip was expected be rough on Long Island Sound due to the weather, deck crewmen skidded about the slippery-planked decks lashing last-minute pieces of cargo to floor shackles. In so doing, seamen tried shielding themselves behind bales of cotton and wood crates hoping to ward off the piercing cold.

Just aft the boiler room compartment was the galley. It appeared to be the warmest place on the ship. Chefs started preparations for cooking the dinner menu. As planned, the food would be served by the stewards at six in the main sitting cabin. The cabin served as a multipurpose room, easily converted by attendants to become the dining area and later the ship's lounge. An announcement by the ship's purser to the passengers informed them that entertainment would follow the evening's meal. Several actors were hired on as part of the ship's complement to play-act with short recitations. Their intent was to make the chilling voyage a memorable event.

At 7:15 p.m., the *Lexington* was steaming along at a respectable 12 knots and positioned four miles north of Huntington Harbors, Eaton's Neck, Long Island. Passenger conversation ebbed so the performers could be heard. Only the ice chunks hitting the hull provided low background thudding noises to the performers' acting routine.

On deck, a crew observer noticed a glow at the base of the midship smokestack. He quickly went to investigate, fighting the blowing snow and wind chill. The lower portion of the stack casing was cracked and on fire. The vessel's forward 13-knot speed fanned the fire, blowing sparks that ignited cotton bales lashed to the deck. Within seconds the flames became uncontrollable. The *Lexington's* mid-ship was burning at a furious rate. Deck crew members rushed to throw buckets of water at the base of the fire. Their efforts were in vain. The fire took a strong hold on the deck.

In the wheelhouse, Captain Child watched in disbelief. His newly commissioned ship was a burning inferno. Another man entering the small cabin reported he overheard Child order his helmsman to turn right and head the *Lexington* for the Long Island shore.

"We'll run her aground at full steam," he cried, pointing toward the black-silhouetted shoreline.

Passengers immediately became panic stricken. Mayhem broke out among the crew as well. With the fear of sinking, everyone darted for the small lifeboats.

The first boat lowered to the water became swamped by the over-crowding of passengers. As the boat rolled on its side, the victims were thrown directly into the icy water of the sound.

The second boat suffered the same fate as the first. Only this time, an over-anxious officer cut the stern davit rope too soon, dumping all aboard into the sea.

Ironically, Captain George Child as well as his first mate and helmsman mysteriously disappeared from the bridge, leaving the ship's command unattended. It was later disclosed at the inquest that these two men were the first to desert the burning ship. In addition, the majority of the crew had little or no regard for the passengers welfare and safety.

Screams seemed to come from all directions in the vicinity of the ship. It was their pleading cry for help. Five survivors later reported, *Only the reflections of the flames offered a small visible indication that the yelling came from persons treading water in the blackened night.*

The ship continued at full steam without a crewman at the helm. She was within two miles of the Long Island shoreline when the intense fire burned through the rudder control assembly. The tiller rope connectors collapsed, causing the ship to turn to port. The ship's reverse direction now blew flames toward the bow of the vessel.

Several individuals untied the bales of cotton, jettisoning them into the water to provide a floating raft. The balance of passengers who perched on the bow jumped into the icy water below. Presumably, these poor souls preferred drowning over being burned alive.

Charleston Hilliard, another of the survivors, vividly recalled how he and his fellow passenger, Benjamin Cox, spent most of the night on floating wood debris.

"We paddled our raft toward the shoreline. Cox, frozen by the icy water, could not continue. He went under without a sound," said Hilliard at the inquest.

The flaming spectacle, now viewed from both sides of the sound only permitted onlookers to stand in awe. Little could be done to help the floundering survivors due to the frigid temperature and high winds. Two small rescue boats were eventually launched from the Long Island coast with hopes of giving aid. Unfortunately, the vessels' high speed prevented any form of rescue. Each boat fought the high swirling seas praying to locate survivors, only to return as they left, empty!

At three a.m., the waters of Long Island Sound decided to finally accept their prey. The *Lexington* sunk beneath the surface of the water just two miles off the inlet to Huntington Harbor. There were no more sounds, no more flames! Just the floating, burned bodies and unrelenting signs of the perilous night.

An immediate investigation was launched to reenact the causes of the disaster. Only five passengers of the likely 118 aboard survived that night of horror. Their testimony, although somewhat contradictory in some instances, was the only evidence of what had happened. Their statements clearly identified that the owners of the ship were at fault, and considered the disaster to be an act of murder!

Later, it was disclosed another ship, the sloop *Improvement*, was less than four miles east of the *Lexington* and saw that she was on fire. The captain, William T. Terrelle, later testified, *"I did not approach the burning ship with my vessel for the fear loosing the tide. I desperately needed it to clear a nearby sand-bar."*

Terrelle was severely criticized by the board of inquiry for his inhumane actions and apparent lack of interest. Since he didn't break any known maritime law, however, Terrelle was discharged without repercussion.

Unanswered questions surfaced. Why did Captain Child desert

his post? Was he the coward that he appeared to be? Why weren't passengers assisted by the crew? Onlookers to the inquest stood in silence waiting for answers that never came.

With the inquest adjourned, little more could be said. The entire event was a travesty. The public again viewed steamboat travel as an unsafe mode of transportation vowing that ships should return to a safer form of sea-travel — sail power!

Remembering the incident in detail, salvagers made several attempts to retrieve what was believed to be large sums of silver from the Harnden safe. One group, headed by Ray Wagner in 1952, achieved success in finding the remains of the *Lexington*. Divers carefully worked their way to the supposed area where the safe was chained to the deck. They found a cart-like object resembling the safe. Unfortunately, the bottom and sides appeared rusted from the seawater. The divers deduced the intense fire somehow weakened the floor of the chest, spilling the silver ingots to the bottom of the sound.

It is also possible that the treasure was extracted from the safe by an unknown salvager and never reported to the authorities.

The keel of the S-5 was laid on December 4, 1917. Commissioned into the Navy on March 6, 1920, displacing 1,088 tons while submerged, 875 tons on the surface. The S-5 length measured 230 feet with a 22 foot beam. Surface speed indicated 15 knots, 11 knots submerged.

Escape From The *S-5*

War was imminent! President Woodrow Wilson recognized the United States' Allied involvement with troubled waters surrounding Great Britain. Eventually, it would cause a conflict unlike any the world had ever seen.

The president and his naval attache realized that safe Atlantic crossings of supply ships would be essential in helping Great Britain. There was an immediate need to assure a constant flow of munitions and equipment — all necessary for ultimate victory in Europe.

Congress became uneasy with the reports of war. Informed sources reported that German U-boats were bring launched in great numbers. At no other time in history was underwater warfare such a threat. The new U-boats had devastating effects on Allied shipping and supplies. Our ships, especially the submarines, were old and outdated compared with Germany's high-tech submarine fleet.

Congress sensed the importance of requests for construction of new submarines to be used by the U.S. Navy. A congressional vote was taken — the manufacture of an astonishing 100 new submarines was quickly approved.

Like all ships, submarine designs are divided into classes. Each class represents some form of design revision over its predecessor. Submarine O, R and S classes were hastily added to the existing obsolete American submarine fleet.

Each cataloged submarine revision provided longer tactical range, greater ordnance, extended underwater endurance and, in general, just bigger, better boats.

Construction contracts were issued to three prime submarine manufacturers: Holland Boat Co., Lake Torpedo Boat Co. and Electric Boat. The companies rivaled each other, each displaying their own unique blueprinted design considerations.

After careful review, the Navy was ultimately persuaded to accept the Electric Boat design program as the ultimate source. Of the three new submarine classes, the S-class was built in the largest quantities. Average length approximated 250 feet and displacement averaged 1,100 tons (submerged).

S-class submarine construction continued well past the end of World War I. At the close of 1925, the U.S. Navy was the proud owner of 51 S-class submarines. Ironically, within the next two years, six boats were lost by freak accidents. The sinking of the *S-5* (SS-110) was the first occurrence of these disasters to become news headlines.

The keel of the *S-5* was laid on December 4, 1917. The boat was officially commissioned into the Navy on March 6, 1920. She displaced 1,088 tons while submerged, 875 tons on the surface. The *S-5* length measured 230 feet with a 22 foot beam. Surface speed indicated 15 knots, 11 knots submerged. Normal crew complemented 4 officers and 37 seamen. Armament consisted of four forward firing 21-inch-diameter torpedo tubes and one aft-mounted four-inch deck cannon for surface confrontation. Marine engineers approved the vessel for a 200-foot dive depth.

After commissioning, the *S-5* completed her shake-down sea trials with flying colors, once again proving American ingenuity. Lieutenant Commander Charles M. Cook, Jr., was awarded command of the new *S-5*. His first mission was to perform a naval recruiting campaign of new officers and enlisted men into the submarine service.

On her first cruise, the *S-5* left Boston harbor. Cook read his orders, which directed him to proceed along the Atlantic coast to the Chesapeake Bay, continuing on to Virginia. The mission ultimately would end in Bermuda, where the submarine with take a northwesterly course returning the *S-5* to Boston. Cook realized he was commanding unseasoned seamen who require considerable sea training in their new boat.

The Sinking

At approximately 50 miles southeast, off the coast of Cape Henlopen, Delaware, Lt. Com. Cook started routine emergency test dives of the *S-5*. He knew, the repetition of dives and surfacing exercises would sharpen the skills of the newly formed crew. The boat's position avoided the heavily transited sea-lanes of the Atlantic shoreline.

On September 1st, Cook sounded the dive alarm. The crew immediately took their respective positions for the descent. Normal procedure for submerging is to flood the ballast tanks, adjust the bow planes and trim ship for the dive.

Chief Gunner's Mate Percy Fox was responsible for operating the forward air intake valve, a critical function to the success of the dive. As the descent continued, Fox was unable to close an open air valve due to a jammed valve lever. As the boat descended, seawater immediately began pouring in through the air ventilator ducts. Fox worked frantically — the valve remained open. Water entered the forward compartments at a fantastic rate. Crewmen reacted immediately, sealing off the forward compartments.

Cook quickly recognized the problem! He tried desperately to bring the *S-5* under control and, ultimately, to the surface. Bilge pumps were activated to full capacity only to discover they were over-taxed as the main gasket seals ruptured. The pumps were deemed of little use. All efforts to regain ascent were lost. The submarine was now sinking bow first. The crew was panic-stricken! The *S-5* settled uncontrolled to the ocean's floor in 170 feet of water.

Commander Cook collected his thoughts, doing his best to assure his men that some form of escape would be secured. At the time, little was known about aqua-lungs or SCUBA survival equipment. Any form of ascent by a seaman from a depth of 170 feet without some form of air assistance was fatal.

Cook deduced that if water in the boat were shifted to the forward compartments, the aft section of the submarine would rise. He knew the boat was 231 feet long. There was the possibility of transferring enough water ballast to the bow, permitting the stern to rise! If the transfer of water were successful, the stern of the boat would be high enough out of the water, to provide some form of escape!

A major consideration was the potential mixing of seawater with the forward acid batteries. Battery cells contained sulfuric acid — mixed with seawater, the combination produces deadly chlorine gas. Cook felt there was little choice. He ordered quick compartment door manipulation on the part of the crew to prevent the toxic fumes from entering the rear compartments.

The Escape

The escape plan was set into motion. With the bow firmly entrenched into the silt, the stern began to ascend slowly at first — the men were joyous. In a split second, the boat's stern began to rotate at a ferocious rate. Men, loose equipment and rushing water rushed forward to the bow compartments. Little could be done to aid the helpless seamen. Miraculously, no one was seriously injured.

The *S-5* settled in an awkward, near vertical position. Her bow was buried in the silt, her stern towering above the ocean's surface. Commander Cook's plan worked!

Inevitably, the feared chlorine gas began seeping into the front compartments. Slowly, the 37 men were forced into the aft compartments of the vessel. There was a ray of hope, however! By tapping on the walls of the hull, they determined that 15 feet of the submarines hull extended out of the water. This proved Cook's original theory.

Due to limited availability of tools, only a small 1/4-inch-diameter hole was hand drilled through the 3/4-inch-thick steel hull. Commander Cook was concerned about the air supply for his men. Under normal circumstances, the boat provided a three-day air supply. With the batteries belching deadly gases there was no way of knowing when the precious air would be exhausted.

Without warning, the power failed, total darkness adding to the fear of the entombed seamen. Four holes were finally drilled — large enough to insert a hacksaw blade. Connecting the holes provided a mere slot of only four inches long and 1/2 inch wide, their only contact with the outside world!

Commander Cook could see ships passing on the distant horizon. He also realized they were too far off to see the stern of his sunken submarine. The work continued. Each man took turns drilling then sawing. The more they cut, the more the edges of their tools became dull and practically impossible to use.

Over 20 hours passed and the hole was enlarged to only a small triangle. The tiny access provided little value to the desperately needed fresh air supply. The opening permitted only the insertion of a pipe with a white undershirt fastened to the end to be used as a possible source of attraction.

Help At Last! The Rescue

The *S.S. Alanthus*, an old freighter making her last voyage to the torches of the scrap yard, passed the ill-fated submarine. Observers on the ship identified the *S-5* as floating debris. They gave little additional thought to the passing flotsam. Rechecking with a pair of field glasses, the ship's captain was confused! Why was there an object moving on the debris? The *Alanthus* performed a 180 degree turn, lowering a boat to investigate the moving white flag. The small boat's crew could not believe their eyes and ears. The *S-5* was discovered at last!

Water and fresh air was immediately pumped into the small opening. Lines were tied around the exposed stern structure protruding out of the water. Since the *Alanthus* was about to be scrapped, she had inoperable radios. She carried little or no tools necessary for cutting the *S-5*'s hull. The captain immediately ordered distress flags to be flown.

Hours went by with little communication with the outside world. Fortunately, the S.S. *George W. Goethals,* passing on a northbound course identified the distress signals flown by the *Alanthus.* Her captain offered on-the-spot assistance. With her radios in full operation, she contacted the U.S. Navy and advised them of the problem.

The Navy put the rescue mission in high gear. Two destroyers, the battleship *Ohio* and several small salvage vessels were quickly dispatched. Each was ordered to supply the necessary assistance to extract the imprisoned seamen of the *S-5*.

Finally, help was on the way. The chief engineer of the S.S. *Goethals* decided to expand the small hull opening. He and his assistants worried that Navy rescue ships were still hours away and it was too long to wait! Once again, the chief's efforts were hampered using only small hand tools. After seven hours of work they did manage to cut an opening large enough to pass the first man into freedom.

Due to the exhausted state of the men, it took over an hour to extract the remaining seamen from the submarine. Lieutenant Commander Cook, following the laws of the sea, was the last man off his ship. Freedom ultimately came to all, 37 hours after the ordeal started.

The Final Voyage

With all the men now safe, the Navy decided to tow the *S-5* to more suitable waters for salvage. The battleship *Ohio* placed a cable around the stern of the submarine, placing the *S-5* in tow. High seas arose from an oncoming storm that engulfed the lame vessel. Despite efforts of the Navy, the tow-line severed, permitting the *S-5's* hull to fill with water. The newly commissioned submarine finally found its resting place on the bottom of the sea.

On August 21, 1921, the *S-5* was stricken from the list of active Navy vessels.

Epilogue

In his final report on the sinking, Commander Cook commented on his crew. He felt each and every man under his command performed their duties with outstanding courage. Every order issued by him and/or his officers was performed to the best of their physical ability. He was proud to serve with men such as these.

Cook continued his career in the Navy, finally achieving the rank of admiral.

Sea Search

Treasure Fleets Of Spain: Where Did They Come From?

Shortly after Christopher Columbus discovered the New World, Spain recognized the unbelievable financial potential that was available in the West. Immediately she made the decision to colonize Mexico and Central America. Within a relatively short period, she received access to South America and eventually the Southwest Pacific marketplace. It was Spain's plan to monopolize all trading between the New World and Europe hence, controlling all world trade and financial flow. Her reign of power started near 1510 and lasted for over 200 years.

The Spanish Crown was clever and cunning. The subjects who had political power were pure-bred Spaniards. Each was well protected by the crown both financially and politically. Every effort was made to provide the homeland with a sound, secure future.

As part of a master plan, Spain colonized the New World and took official control of her newly formed possessions. Colonists were not permitted to acquire trade goods from the European countries unless the merchandise was sent from Spain and transported by ships flying the Spanish flag.

Government officials decided to send merchant convoys twice annually from Spain to the New World. Each ship's hold was filled with European cargo to feed the colonists with essential supplies not available in the Americas or the Southwest Pacific.

Return voyages carried vast sums of gold, silver and precious gems supposedly traded for European goods. It was their theory that once the Spaniards controlled the Philippines, Oriental silks, rugs and porcelain-ware (items unheard of prior to this time) would become prize possessions for the European wealthy. Manufactured goods sold at a rapid pace. Europeans quickly became envious of the successful Spanish trade. Other neighboring countries England, France and Holland wanted their share of the New World. Spain was no fool: she insisted on her monopoly!

The value of the New World trade became more exciting with the arrival of each returning ship. Thoughts of riches tantalized the minds of government officials as vessels emptied their holds. Large financial returns were guaranteed by saleable goods derived from the New World.

In addition to the King's own greed, the Spanish government continually increased its financial debt to other European countries. The Spanish economy demonstrated serious instability, demanding an exponential need for riches and cash flow. Despite her debt, Spain remained a dominant power throughout the 16th century and would retain her world position as long as treasure ships brought back their riches.

Her lack of sound monetary planning provided very poor future investments. Fortunes were recklessly spent for unjustified military ventures. As a country, Spain lacked the resources to support any form of military power other than its New World gold and silver. Spanish wealth stretched halfway around the globe to the Southwest Pacific. During the estimated 250 years of her reign, Spain looted $40-50 billion from the West! She achieved controlled dominance of over 200 of New World cities. However deceitful her plan was to other European countries, it did appear to be working.

The Spanish became *the devils of their day*. Very little value was placed on human life or limb. They destroyed the Peruvian Inca and the Mexican Aztec empires. A million-plus Indians were slaughtered in Panama. Natives of the West Indies actually worked to their death. Labor death rates became so high the Spanish lords forcibly imported black slaves from Africa to make up for the lost work force. The majority of today's population of the Caribbean Islands are the direct descendants of these impoverished African slaves.

Native laborers were needed to work in the gold and silver mines throughout the New World. In general, animals were considered more precious than humans. Spanish overseers demanded greater daily output from an Indian worker than they did from a mule! When laborers complained or did not take direction, they was unmercifully flogged. If an individual survived his beating, he was forced to continue working or left to die. The next man in line was expected to take up the dying man's workload.

Vast new discoveries of silver necessitated other methods of moving the precious ore. The Spanish decided to mint New World coins of their own denomination. Coinage made for easier transportation and provided a more accountable solution to the actual amount of gold and silver mined.

Mint coin production methods were relatively simple. Pure silver rods were poured to the exact size of the coin. The ends of the rod were sliced off, providing a coin blank. Each blank was struck with a hand die that left an impression onto the soft metal. The finished coin was called a *Cob*. A name derived from the phrase *Cabo de Barra* or *cut from a bar.* The entire minting operation was performed by hand. Often coins were struck off center, accounting for their unbalanced appearance. Several *Cob* designs minted followed the old Italian Renaissance masters such as Benvenuto and da Vinci. Originally, *Cobs* were made only from pure silver. Gold was shipped in ingot form. The minting process of gold coins started later in the 1600s.

Sometime during 1513, Spain created a new governing agency, *Casa de Constrastacion*. A much needed governing body designed to oversee the New World trade. The *Casa* served as a school for up-to-date navigation training and provided an official record-keeping operation for the crown. Vessels that sailed had an appointed record keeper, the *Escribano*. His duty was to record officially all cargo that loaded and unloaded onto the ship. Official manifests became a record for collecting the king's tax. Spain placed a whopping 20% tax levy on all New World trade! She required large amounts of tax revenue to support her thirsty need for cash flow.

It wasn't long before the king was convinced that his country was being robbed of a fortune! He recognized many of his trusted employees were actually *tapping the till,* or stealing from the crown, for their own advantage. The fire was calling the kettle black!

The crown finally imposed new, strict laws against smugglers, pirates and thieves. Convicted felons immediately became sentenced to five years as galley slaves. The expected lifespan of these unfortunates was two years for the physically weak, four years for those with stronger backs and minds. Most of the convicted prisoners preferred death over pulling the oars and being beaten by the galleon master's whip. Very few survived the full sentence of their cruel incarceration.

Pirates, on the other hand, once caught, were sentenced to an immediate hanging, always considered a quicker and more humane way to die. They were the gallant swashbuckling thieves of their day.

Despite the king's severe penalties, smuggling and thievery continued. An estimated 20 to 25% of the precious treasure smuggled into Spain was never taxed.

It became necessary for the Spanish to reduce their losses and cut the threat of piracy. The government opted to organized merchant ships in convoys. These strings of ships were called *flotas*. The Spanish Admiralty decreed all merchant ships sail in *flotas* and run alongside protective galleons. Often *flotas* would range in size from 10 to 40 ships per sailing.

Convoy structure consisted of a lead galleon, the *capitina*. This ship carried the commanding officer of the *flotas*. He was the officer responsible for the overall success of the mission. He handled all navigation skills and provided decisions on weather observations. Other ships in the convoy relied on his judgment to assure their safety and well-being. The *almiranta* was the ship protecting the rear of the convoy. Her sole responsibility was to keep a constant surveillance for an unsuspecting attack from the stern by a renegade pirate ship. She carried an officer who usually held the rank of admiral. The Spanish admiralty delegated this officer as second in command. In the event the commanding officer of the *capitina* was taken ill or killed, this man would assume command of the lead ship. It became the admiral's responsibility to assume command of the entire convoy in the event of a sea battle. The chain of command continued to the leader of the on-board soldiers. This officer was given the title of *gobernador*. His ship was designated the *gobierno*.

Spanish galleons were small by today's standards. Most were only 100 feet long. Hull designs were poor and difficult to maneuver. When confronted with bad weather, each ship became a floating disaster. The typical sterncastle sat 40 feet above the waterline. Standard mid-ship design displayed two masts; larger vessels consisted of three. The ship's cumbersome forecastle was a mariner's hindrance to visibility and added to the poor speed performance of the craft.

A seaman's life was barely tolerable, at times, impossible! Sanitation was virtually nonexistent. A community bucket was used by deck hands for human waste. Water and food rations generally turned rancid. Ship smells became nauseating stemming from live animals, human waste and vomit. Lice and filth prevailed everywhere. Below decks, seamen slept in hammocks slung between foul-smelling cannons. The ship's surroundings were uncomfortable, hot and dismal. Every voyage reported cases of rickets and scurvy. Once a seaman contacted any form of illness he was placed in quarantine for fear of plague. Fast recovery was imperative. Often the remedy and/or solution to an illness resulted in feeding the poor soul to the sharks.

Passengers and officers were only slightly better off. They had the comfort of living in the sterncastle of the vessel. Their cramped quarters were considered luxurious by the standards of the day. A servant was available to care for the food and to supply whatever comforts deemed necessary. They too had to endure the ship's other traveling companions: rats, lice and bedbugs and foul smells.

Galleons were considerably slower than the pirate sloops and brigantines. Pirates demanded speed and maneuverability from their ships and crew. First order was to outrun the heavy guns of the galleons. A top-heavy galleon relied on its ability to provide great firepower once a target was brought into range. Any opponent caught in crossfire was considered a quick victory for the heavy warship's commander. A surprise attack from the galleon's stern was the enviable position for unscrupulous pirates. The awkward, square-rigged sail design limited the galleon's ability to fight efficiently since it was nearly impossible to outmaneuver her enemy.

Recognition of others at sea was often difficult! The Spanish masters found it hard to distinguish between renegade pirates and another government vessels. For the most part, once another ship was sighted, it undoubtedly meant a fight!

England, Holland and France became tired of the longtime Spanish supremacy. The leading countries of Europe wanted to break the Spanish hold on the valuable ocean trade. Often, their naval officers looked the other way when it came to dealing with pirates.

Pirates believed the old axiom, *a dead man tells no lies.* Upon capture of unsuspecting prey, they confiscated the treasure and ruthlessly eliminated the competition. Pirates of the day had high respect for gallantry and bravery. Little value was placed on fancy-outfitted officers and high-ranking officials, however! Assuredly, a pirates' victim died a slow torturous death. Pirates often commented, that their Spanish enemies taught the world these cruel and inhuman ways.

Every sailing vessel respected bad weather. Hurricanes were in a special class by themselves! One of the most famous natural fleet disasters was the loss of the 1715 Fleet or Plate Fleet. It consisted of a 12 ship convoy that became the focal point of ships returning from South America, the Far East (The Orient) and Mexican ports.

All ships rendezvoused in Havana to form a flotilla for return to Spain. Each vessel carried a large portion of the king's ransom of gold, silver and precious silks. The time of departure ran into the hurricane season during the summer of 1715. The captains' were not sensitive to the potentially violent storms frequenting the Florida coast.

Finally, a departure date and course were set. All 12 ships were to return to Spain via a northerly flow of the Gulf Stream hugging the East Coast, then upon reaching the Carolinas turning east toward Europe.

With all ships underway, a fast-moving tropical hurricane moved across their path of sail. According to the plan, each of the treasure ships was responsible for her own survival. The fearsome hurricane of 1715 sank all but one ship. Several vessels blew toward land, others sunk off the nearby Central Florida coast in deep water. Reportedly, each ship carried a large fortune in gold and silver bars. Hundreds of the crew lost their lives. Ironically, there were a few survivors who unbelievably swam to the mainland at St. Augustine, Florida. The surviving few found refuge at a Spanish fort and awaited help from Spain.

Of the original 12, only one ship survived the storm. It fought its way back to Spain to tell of the misfortune. Government officials were skeptical of their story! However, the admiralty decided to start salvage operations immediately, dispatching two ships to Florida.

Aided by local Florida Indian divers, the salvage group yielded only a small amount of the crown's fortune. The majority of the treasure remained in water too deep for the breath-holding divers. Spain continued to search for several years thereafter. Only tiny portions of the treasure were ever brought to the surface. Frustrated, Spain eventually abandoned the project.

Historians say, that during the 200 to 250-year period the Spanish ruled the seas, as much as 20% of their treasure fleet never returned to the crown. Their fortunes are still either hidden in remote treasure troves or buried beneath the oceans they sailed.

The ultimate achievement of success was clearly shown later in history by other ruling governments in the Caribbean: the Dutch in Curacao, English in Jamaica, Bahamas and the Lesser Antilles, ultimately, the French in Martinique and Haiti and finally the Portuguese in the new penal colony of Brazil.

The *President Coolidge*

The *President Coolidge* was built by the Donner Steam Ship Line in the late 1920's. She was 654 feet long, weighted 22,000 tons, and cost $8 million to construct. Launched in 1931, the ship was named after the 30th president of the United States and was christened by his wife, Mrs. Calvin Coolidge.

Air travel had yet to become a feasible means of trans-Pacific travel in the early 1930's. The Donner line built the *President Coolidge* to supply a fast means of travel between the United States and the Orient. For almost 10 years she routinely made the crossing from her home port in San Francisco to the Philippines, Hong Kong and Japan. On her maiden voyage, the *Coolidge* carried 1,200 passengers across the Pacific in what was then a record setting-time of 12 days.

The bombing of Pearl Harbor in 1941 brought the United States into the war. Thousands of troops and support equipment needed transportation to the South Pacific. Like many other ships of her time, the *Coolidge* was stripped of most of her luxuries and converted into a troop transport to carry large quantities of men to the war zone.

Entering the ship, troops admired the palatial staterooms that once offered fine silver furnishing and linens. These quarters normally carried the rich and famous and were now reduced to high-volume tenements housing rack upon rack of bunkbeds. Still in place were the swimming pool and sitting salon displaying an ornamental fireplace. Above its mantel sat *the lady*, a statue of a woman standing in front of a horse. Several plush rooms were converted into rows of sea-flushing toilets to accommodate the large complement of troops. Even officers were forced to live in tight confinement until the ship reached its normally untold destination.

On the morning of October 24, 1942, the luxury liner was sailing near the island of Espiritu Santo, off the Truk Lagoon. In her new capacity, a troop transport, she had 5,150 men on board.

In command was Captain Harry Nelson, a seasoned skipper who was known for his skillful seafaring ability. The *Coolidge* turned to enter the Scorff Channel, a 1.25 mile passageway between Turtle and Bogacio Islands. The inlet was strategically loaded with floating mines to ward off invading Japanese submarines.

As the ship entered the channel leading to the harbor, Captain Nelson stood watch on the bridge awaiting signals from the shore. A worried-looking seaman darted onto the bridge carrying a message that had just come over the radio from the Allied troops on shore. It read,

"DANGER...YOU ARE ENTERING A MINE FIELD. STOP! REVERSE COURSE IMMEDIATELY!"

The captain no sooner finished reading the message when the bow of the ship was rocked by an explosion. The captain gave the order, "All engines, full astern!"

The first mate pulled the levers on the bridge telegraph signaling the engine room, but the ship was too slow to react to avoid hitting a second mine.

Solders and crewmen scurried about the deck not knowing what happened. Below decks, there was mayhem with the men stuffed into small confined spaces. Everyone realized there has been an explosion of sorts, but no one had any idea as to how severe the damage was!

Within minutes, the *Coolidge* began listing to port. The captain realized his vessel was in serious trouble and about to sink. Nelson then ordered his helmsman to turn the ship toward the island and run her aground. Since there was no time to check charts, the ship ran aground on a sand bar. Shaking his head in dismay, he was forced to gave another order, "Sound the alarm, abandon ship!"

Deck crewmen reacted quickly, tossing huge rope nets over the side to provide an escape route for the stricken solders of the 172nd Infantry and crew.

Having a complement of over 5,000 men, evacuation of the ship was slow. Someone called out, "Don't panic, we have run aground. We'll all get off okay!" Most of the crew didn't believe the ship was in danger of sinking.

The explosion caused the ship's forward oil tank to rupture. Water surrounding the hull became saturated with a heavy oil slick. Solders clinging to the bottom of the side nets feared the oil would catch fire and they would be caught in the inferno as they groped their way to shore.

Fortunately, small landing craft, LCTs and LCVPs surrounded the ship and were able to offer immediate assistance in getting the men off the ship.

Within an hour after hitting the mines, the *Coolidge* listed dangerously. The shift in the ship's attitude caused the base of her hull to slip off the sand bar and slide into water exceeding 240 feet were the danger of sinking became a sudden reality.

Orders were given by officers above to push anyone off who was found hanging on the bottom of the evacuation nets. Men waited patiently as each took his turn to descend the ropes. Slowly, the men fought their way to the shoreline for safety with the aid of the small boats.

Another group of solders still under deck became trapped in the ship's galley. Unable to escape due to the listing and onrushing seawater, they would have perished if it had not been for the heroic efforts of Army Captain, Elwood J. Euart.

Captain Euart tied a rope around his waist and lowered himself to the trapped men through a sea wall within the ship's hull. After helping each man to the rope were they were able to climb to freedom, Captain Euart attempted to climb out himself. Sadly, the heroic officer became too exhausted to save himself and went down with the ship before he could be helped.

Miraculously, they're were only two fatalities when the *Coolidge* went down: Euart and Fireman Robert Ried, who died from the explosion when the ship hit the first mine.

Two hours after hitting the first explosion, at approximately 10:30, the *Coolidge* slipped beneath the waves.

The sinking of the ship by our own mines was a great embarrassment to the Navy. A complete investigation was set into motion. No one, including the captain was ever charged with the unfortunate mishap.

Despite the disaster, Captain Nelson was later made commodore of the American President Lines and continued serving as master throughout the war. After spending 53 years at sea, Nelson retired in 1948.

Diving The Coolidge

At 654 feet, the *President Coolidge* is 150 feet longer than the biggest ship in Truk Lagoon (*The Shinkoku Maru*). This makes it the largest shipwreck in the world within sport diving depths. Bigger wrecks such as *Titanic* do exist, but divers using only scuba-type equipment cannot reach its enormous depth.

The *Coolidge* can be reached from either shore or boat. Two lines lead a diver to the bow in 70 feet of water. One runs from a buoy on the surface, the other from a decompression/safety stop area in 15 feet of water just off shore.

Because of the depths involved on the *Coolidge*, it's a good idea to ascend along the line leading to the safety stop area. A hang on bar is anchored with cement where the line ends. This arrangement has proven to be very convenient.

The ship displays two forward cargo holds, both filled with tires, jeeps, large cables and piles of other automotive debris. Depth in the forward hold is about 100 feet.

The *Coolidge* lies on her port (left) side. When swimming the wreck, it's helpful to acclimate oneself to the ship's position. I found a large artillery gun outside the second cargo hold that was almost missed because my perspective was off.

The *President Coolidge*

On the right side of the hull near the bow, there is a fairly large gun mounted to the deck. Just below the gun, I consistently found a group of scorpion fish. Certain areas of the wreck are home to high populations of these unusual inhabitants.

Beyond the second cargo hold lies the towering superstructure. Entering on the shallower right side of the hull, one can swim down a long hallway and still have large areas of open water overhead. Rifles, helmets and other military debris can be found in this corridor.

After swimming two-thirds of the way down the hull your guide (dive with a guide or you'll never find these relics.) will lead you deeper inside the wreck to a large room with a fireplace. Above the mantle sits *the lady*, aforementioned, the statue of a woman standing in front of a horse. Depth at *the lady* is 160 feet. Do not attempt this dive without a guide!

Above and behind *the lady* in 140 feet of water lies the rows of toilets that were put in when the *Coolidge* was converted to a troop carrier. Swimming beyond the toilets takes you past a tiled fountain, and back out the front of the superstructure just below where you entered.

It would take weeks of diving to see all the *Coolidge* has to offer. In addition to *the lady*, the swimming pool and soda fountain are also worth seeing.

This gallant relic of World War II is a must-see if you're visiting the Truk Lagoon. The area is filled with historic Japanese ships that tried to fight off Allied forces and eventually found their demise.

This may be the last photo taken of the Morro Castle *before she found her watery grave off the New Jersey shore. The ship was considered a masterpiece of design and construction — the envy of her competition.*

Morro Castle, Ship Of Doom!

The depression years were in full swing. The country's morale was at an all-time low. Jobs were scarce, pay generally poor. The lucky working few were thankful for the little they had! Surplus vacation money was difficult to save due to the poor financial state of the nation.

The *Morro Castle*, a luxury ocean liner owned by Ward Lines steamed her way into Havana harbor, dropped anchor and filled her holds with provisions for the return trip to New York. The four-year old, 11,500-ton liner, commanded by Captain Robert Wilmott, was the pride of the company.

Launched on her maiden voyage in 1930, she was considered the latest in design and offered the ultimate in comfort for her passengers and crew of 231. Naval architects made every provision for safety, including nine watertight compartments, electrically operated bulkhead doors, ample fire hydrants and steel-floored deck plates in lieu of conventional wood planking. Attached to her davits were twelve 37-foot lifeboats constructed of fire-resistant steel. Lifeboats were capable of carrying 65 passengers and/or crew. The *Morro Castle* was considered by some as the new *Titanic*.

Prior to departure, Captain Wilmott was informed of unrest among the crew. They complained of poor working conditions including long hours and low daily wages to name a few. Deception prevailed everywhere. Later, it's reported, some of the passengers suspected something wrong! Upon arrival in New York, Wilmott

Sea Search

planned to meet with his superiors to discuss the seriousness of the labor problem.

Aboard were the chosen few who proved to be unaffected by the depression — 318 suntanned passengers returning from a holiday cruise originating only eight days prior.

The captain weighed anchor, departing per the published schedule. Those aboard hoped for smooth seas, but the September overcast provided a threat to weather — this was the beginning of the East Coast hurricane season.

Undertones of crew complaints persisted — engine room, stewards alike, especially with the ship's radio operators, George Alagna and George Rogers. The captain had had a previous run-in with Rogers' conduct before. He felt the radio-man required constant watching. The captain's thoughts eventually returned to his duties, managing the ship.

The second day after departure from Havana, an unexplained fire broke out in one of the ship's holds. Several of the service personnel were able to extinguish the flames. The event was tactfully kept from passengers. However, a few did make casual mention of detecting smoke in the companionway of the ship.

Captain Wilmott initiated an investigation. Reports were contradictory. It was impossible to determine the cause of the fire or identify a motive. Wilmott was confident the incident was caused by arson although he could prove nothing! This event added to the captain's fears of the crew's attitude and discontent. Was the labor problem festering?

The ship finally seemed as though it was settling down. Passengers appeared jovial, normal ship activities continued without incident. The luxury liner proceeding uneventfully along its prescribed northbound route. Visibility permitting, land could be seen on the horizon.

On the evening of Friday, September 7, 1934, the following series of events occurred, becoming some of the most bizarre activities ever reported in maritime logs.

It's the practice of most ships to have passengers attend a black-tie cocktail party followed by a formal dinner with the captain of the ship. Festivities aboard the *Morro Castle* were arranged for the last night at sea to celebrate a final farewell to the close of the voyage.

Passengers became irritated with the delays in serving the night's meal. Several individuals expressed disappointment that Captain Wilmott was not at their table! It was explained that the captain was unavoidably detained. Others waited patiently as music played effortlessly in the background.

Without an explanation, a crewman entered the dining room with tears in his eyes. Shaking his head, his voice cracked as he tried to speak, "The captain is dead!"

The entire dining room was stunned by the words of the sailor. Low-toned comments whispered through out the dining salon. "What happened? How did he die? Who's in charge of the ship?"

Startled by the sad news, passengers and crew became confused and terribly upset. The ship's social activities were quickly cancelled. Gloom was felt by all. The *Morro Castle* was now only hours from docking at her berth in New York.

William Warms was second in command under the late Captain Wilmott. The chain of command dictated that Warms take charge of the liner. Although a seasoned veteran of the sea, well educated by prior experience, he has never commanded his own vessel. The new captain was nervous and bewildered by Wilmott's death as he too had no explanation!

The ship's doctor, DeWitt Von Ziele, stated his first observations indicated a possible heart attack! He also commented on the possibility that Wilmott may have died from some form of poisoning! Or, was it perhaps murder! There was no way of identifying the actual cause of death until an official autopsy could be performed in New York.

Acting Captain, Warms took his place on the bridge of his new command. The ship's course now approached the shoreline of New Jersey. Making observations from the bridge, seas were building and visibility worsened as rain showers began to intensify.

A tense feeling of uneasiness emanated from the men on the

bridge. The captain's death had everyone on edge. An unexpected call came from the interphone. One of the ship's crew informed the bridge, "There is a distinct aroma of smoke coming from a storage locker!"

Captain Warms became suspicious. Was this a coincidence or an other arson attempt like the hold fire of just a few days ago?

Another report came to the bridge confirming the first observation, and the passengers soon became aware that there was something wrong. The fire was spreading.

Warms finally sounded the fire alarm. The ship's stewards ran up and down companionways banging on cabin doors to alert the drowsy passengers. It was 2:30 and most of the unsuspecting passengers were asleep. Several appeared intoxicated from the party begun the night before.

The fire became more intense and smoke filled the passageways. Passengers choked from the toxic odors and several portholes were broken to relieve cabins of smoke. Instead, the portholes acted as ventures, causing the fire to be fanned with the high on-rush of air.

The ship was still operating at normal cruise speed. Warms failed to issue any signals to the engine room for reduction of speed. The ship's forward motion mixed with high winds from the storm aided the fury of the flames.

The new captain gave the command to close the watertight compartments in order to prevent the fire from spreading. Crewmen refused the order, fearing passengers would be trapped in the sealed compartments below decks.

Warms still did not transmit the international distress signal, S-O-S! Radioman George Rogers did nothing. He awaited the formal command from the captain. Captain Warms was so preoccupied with the confusion, that he completely forgot to send the dispatch advising other ships at sea of their dilemma.

During the haste and commotion, crewmen left several of the fire hoses open below decks. Water suddenly poured into hallways and down stairwells aimlessly. Open, unattended hydrants reduced the necessary water pressure above to fight the fires. Sprinkler systems too were of no value! Several of the freshly awakened passengers became totally disorientated, wandering up and down passageways, adding to the confusion.

Although still dazed, Captain Warms collected his thoughts, finally requesting the S-O-S be transmitted. Radioman George Rogers, started keying the transmission as ordered. It was now 3:10. Then, without warning, the ship's main generators stopped. The *Morro Castle* fell into total darkness. The only light available was from the flames themselves! Further transmission was impossible, there was no power to activate the radios.

Several miles to the east of the *Morro Castle*, a freighter, S.S. *Andrea S. Luckenbach,* ran a parallel course to the *Morro Castle*. Observers on the bridge of the freighter noticed the intense glow on the horizon from the burning vessel. Their radio-room made no mention of a distress call from any ship up to this point. In spite of the radio silence, the *Luckenbach's* captain ordered a turn toward the burning ship to offer assistance. He was certain something was wrong!

Passengers, looking for commands by the officers and crew, started jumping into the chilling September water. Their only hope for survival — swim to shore, use the dim lights of the New Jersey coast as a beacon. Driving rain and crested seas persisted, adding to the difficult task of swimming the long distance to safety.

The drudgery of lowering lifeboats became unmanageable. Block-lines, attached to the ship's davits, were difficult to release due to the forward speed of the parent ship. Several lifeboats were found attached by only one block! The small craft capsized, throwing unsuspecting survivors into the churning sea. The boats were precariously dragged alongside the *Morro Castle's* hull.

Panic spread among crew members. Partially filled boats made hastily departures from the burning ship. In their haste, lifeboats failed to stop to pick up pleading survivors from the icy water.

The *Luckenbach*, now fully aware of the disaster, broadcast the necessary S-O-S. The cry for help was quickly responded to by the

Coast Guard, local fishing boats and several small pleasure boats, each offering aid to the floundering survivors. Visibility was severely hampered by heavy rain and churning seas. Passengers were detected by water reflections of the burning ship.

"It seemed there were bodies everywhere," reported one seaman from the *Luckenbach*.

As twilight broke over the horizon, search vessels were able to identify the few unfortunate survivors still bobbing in the water. The remaining souls, displayed weakness from exhaustion and exposure due to the cold night water of the Atlantic. Nothing could be done for the dead bodies floating among the cast-off debris.

On the bridge, the ship's clock chimed four bells, indicating to Captain Warm and his remaining 14 man crew that it was 10:00 Saturday. The *Morro Castle* was still burning out of control. Little could be done now to aid the doomed ship.

The Coast Guard cutter *Tampa* came alongside the lame *Morro Castle*. In command was her skipper, Lieutenant Commander E. G. Rose. The *Tampa* attempted to take the *Morro Castle* in tow with hopes of salvaging whatever was left of the burning ship. The sea was still rough from the persisting storm that laid off the coast of New Jersey.

Within minutes, *Tampa's* propellers became fouled in the tow line stretched between the two vessels. Commander Rose feared she too would catch on fire from the still-burning *Morro Castle*. Rose issued the order to expel the crew and cut the towline. The *Tampa* dropped anchor, permitting the *Morro Castle* to float toward the Asbury Park Beach, totally unaided. There was little else to be done. Finally, the burning liner came to a halt at the water's edge.

Thousand of onlookers gathered on the beach to stare at the smoldering devastation, all in awe of the one-time ship of luxury. Newspapers reported, *There was a carnival-like atmosphere. Where else could you get such a show, free of charge?*

Final Conclusions

The board of inquiry found Captain Warms guilty of poor

directional commands to his crew and passengers. He was charged and found guilty of gross negligence. Additionally, they commented, "The ship should have been slowed to minimize the draft that helped fan the blaze." Warms was sentenced to two years in jail.

The premature evacuation of the engine room personnel commanded by Chief Engineering Officer Eben Abbott, caused short circuits in the main generators. Had Abbott stayed tending to his duties, electric lights would have remained on to aid in the abandonment of the ship. His cowardice would cost him four years behind bars.

Warms was to serve his prison term. Abbott served only three years of his sentence. Later, the Federal Court of Appeal proclaimed, it was the poor discipline of the men demonstrated by the late Captain Wilmott that caused the discontent among the crew. Charges were later dropped against both men.

The agitating radio operator, George Alagna, was never charged with any of the crimes. He did carry the stigma of a suspected arsonist by several of his peers. Was he the one who caused the doom of the *Morro Castle*?

Alagna attempted suicide. Failing, he was placed in an insane asylum for treatment. After release, Alagna attempted to murder his partner for trying to steal a job from him. He was sentenced to four years in prison for the dreadful act. Investigation disclosed that Alagna had a previous police record. During his youth, he was formally accused of child molestation, theft and arson.

Time passed for George Alagna. In 1953, his life ran into another turmoil, when he was found guilty of murdering a farther and daughter to whom he owed money. The judge passed sentence, this time issuing two consecutive life terms in prison. Alagna died of a heart attack in the penitentiary 12 years later.

There are still several unanswered questions to this mystery. First, how did the two fires aboard the *Morro Castle* start? Captain Wilmott's body was later found and reported as charred remains. The autopsy was never performed.

Next, did he die of natural causes or was he murdered?

Although George Alagna never admitted setting fire to the ill-fated ship, was he the guilty perpetrator?

And finally, were all of the unfortunate circumstances strictly coincidental or do the answers lie with the 134 lives that perished during that night of terror?

Oak Island, The Home Of Pirate Treasure!

On the north central coast of Nova Scotia, Canada, lies a small pine-covered island noted on the map as Oak Island. It measures a mere one mile long and half mile wide. Toward the northeastern quadrant of the island rests a body of water identified as Smith's Cove. This small bay with its natural surroundings has excellent protection from the elements and visual detection from the sea. The insignificant plot of land may represent the most interesting treasure site ever dug by a pirate! The varied effort of numerous salvagers throughout the centuries to find buried treasure has caused more unsuccessful attempts to locate a believed fortune than any other recorded to date.

During the 15th and 16th centuries, Spanish conquistadors discovered New World riches of gold and silver in Central Mexico. The greedy demands of the Spanish government perpetrated the need of returning these riches — hence, treasure ships were formed to bring home the newly found fortunes.

Galleons loaded with the valuable cargo filled their holds with supplies and set sail for Spain. The king impatiently awaited the sight of the needed riches to satisfy the country's thirst for money.

Cuba became a necessary mid-point stop where fresh provisions were added to the ships and then flota convoys were formed to assure their safety from pirate warriors.

The voyage from Havana harbor required a northerly heading following the Gulf Stream along Florida's coastline. Once along the

Carolinas, vessels turned east, making a direct run for Spain. This portion of the trip was expected to take another two to three months of sailing.

The shoreline coves of Cuba and its neighboring Caribbean Islands, granted refuge for unscrupulous pirate ships that patiently waited for the heavily laden galleons. Swashbuckling buccaneers salivated with the thoughts of the riches carried in their holds. The treacherous open seas offered pirate ships the opportunity to raid the slow-moving galleons before they reached Spain.

With their flag of skull and crossbones flying at high mast, fast-moving pirate ships attacked unsuspecting treasure galleons — usually from the rear — stole their booty and ran for the safety of cover in an uncharted harbor. These fearsome acts had familiar repetition.

Some speculate, that the famous Captain Kidd performed such hijackings many times. It's theorized, on one occasion, he discovered a slow-moving galleon sailing alone northbound. The ship's low-slung silhouetted profile qualified the vessel as a treasure galleon returning to the coffers of the king. The ruthless pirate overtook the galleon, murdered the crew, stole the loot, then sent the ill-fated galleon to the bottom in flames. Instead of returning to the customary safety of the Caribbean, Kidd headed north to Nova Scotia — a new land for pirate cover, a course other smugglers would eventually follow.

Often the booty was divided among the crew, with the largest share going to the deep pockets of the captain. Kidd was no exception. He realized it was impossible to carry so much gold, silver and jewels aboard ship. He, along with other peers of his profession, preferred to bury the treasure and retrieve it later.

It's thought Captain Kidd, with his team of dubious looters, had to deposit the booty in a safe, well-protected area. Oak Island may have been the site of that burial! Several artifacts later uncovered from "the pit" indicated the treasure stemmed from Spanish origin.

There was later speculation that most of Captain Kidd's loot was eventually accounted for and added doubt that it was he who designed and dug the pit.

Sir Frances Drake sailed under the auspices of Queen Elizabeth I of England. With her consent, Drake found her legalized form of piracy justified! English ships were fast and elusive. Spanish settlements within the New World had poor fortifications, usually manned by Spanish renegades. The small villages and the treasure ships anchored in their harbors became fare game for attack by Drake's seasoned, well-trained seamen.

Although record keeping was minimal, much of Drake's life was documented. There were suspicious periods of his voyages during the mid 1570s however, that are unaccounted for! It's possible that Drake was overwhelmed by the fortunes he and his men pilfered. Later, Sir Frances may have decided to keep the treasures for himself! Was Drake the possible pirate who masterminded the cache buried at Oak Island?

Other pirate names associated with the Oak Island mystery are possibly Henry Morgan, Blackbeard the Pirate (Edward Teach) and William Phipps. Of all these scoundrels, Phipps was the most likely candidate. He originally hailed from Maine and sailed the coastal waters of New England and Nova Scotia. Documentation shows the man was not adverse to pillaging and plundering French villages along the Canadian coast. One settlement in particular, LaHave, was only 25 miles from the Oak Island site.

Historians question the validity of these men sailing 2,000 or more miles from the warm waters of the South to bury their treasures. The few relics uncovered did prove that someone made the effort to build a well-engineered fortification for their booty.

The identity of the individual(s) who actually engineered "the pit" is of little consequence, weather it be Captain Kidd, Sir Francis Drake or someone else. Some psychics speculated that an unknown robber secretly buried the Holy Grail in "the pit" along with the missing crown jewels of France!

The name of the culprit only adds credibility to the value of the potential treasure. It is clear that whoever it was had the clever, ingenious ability to deceive anyone trying to uncover the secret.

The Treasure Site

"The pit," as it was later called, was found accidently in 1795 by a teenage boy, Dan McGinnis. McGinnis was told by local residents that the island was the resting place for several pirates who had frequented the area years prior. Thinking it may have been a possible buried treasure site, the young man employed the aid of two other friends to help search for a possible find!

After digging only two feet below the surface, the boys found several layers of carefully laid flagstones. Their excitement grew, they thought they had found a real treasure.

Dan McGinnis and his friends continued to scoop out their find. At the 10 foot level, an oak log platform was uncovered. The hole by this time had grown to 13 feet in diameter.

Confident now they were onto something, the young men persisted, shoveling to the 20 foot level, they unearthed another oak platform. Upon reaching 30 feet, the next platform was uncovered. This venture was beyond their greatest expectations. McGinnis knew this was too much for them to handle alone. He was forced to seek assistance.

McGinnis went to his friend John Smith. After showing Smith the impressive evidence , he convinced Smith to buy the surrounding property. Together, they farmed the area to raise money so the excavation could continue. In their spare time, both men worked "the pit" with the certainty of reaching the ultimate cache. The frustrated excavators eventually exhausted all of their limited private resources.

In 1802, another partner, the Onslow Company was added to the roster for continued financial support. The shaft now reached a depth of 90 feet. Platforms of oak covered with pitch and stone were continually uncovered. Whoever engineered "the pit" was considered a genius!

A large stone weighing almost 200 pounds was found resting on still another set of oak timbers. The elongated stone had inscribed markings representing a form of cipher. Were they the written directions for the salvages to follow? The limited expertise of the workmen provided little help in determining the meaning of the authored clues. The stone was later reviewed by many onlookers with hopes of finding a key that would unlock the mystery.

Viewers provided no answers, only adding to the confusion of pit crews. The stone ironically disappeared many years later. The inscriptions may have been the needed ciphered instructions that would lead someone to the treasure.

The work continued. At 98 feet, still another oak platform appeared — surrounded by a cement like substance. Without warning, unexplained leakage of seawater appeared from the wall, quickly filling the shaft. By the following morning, the pit was filled to the 60 foot level. It was later determined, the deluge of water came from a man-made flood tunnel originating in the direction of Smith's Cove.

More determined now, the Onslow group dug another nearby shaft to reduce the threat of water seepage. At 110 feet, the shaft took a turn, digging horizontally to intercept the original shaft. Workers planned to undercut the first pit with hopes of finally finding the expected treasure. Assuming the 10 foot repetition of platforms would continue, the idea made good sense. This theory was surly the solution.

Digging continued — as before, the sudden seepage of water appeared. This time, workmen fought for their lives. Only the rapid exodus of the excavation party from the second pit avoided a disaster.

The unexplained second flood quickly dampened the interest of existing and future investors within the Onslow Company. Excavation was forced to cease due to their financial dilemma. Interest in the treasure site lay dormant for the next 45 years.

By 1849, a new group of investors decided to pick up where the Onslow Company left off. The Truro Company was formed by a group of prominent businessmen who felt the site was worthy of their investment. Prior reports of site findings and samples were supplied to the new investors, justifying the continuation of the search for the lost treasure. It was their intention to take more aggressive action with the now called "money pit" than the previous treasure hunters.

The new excavating team's first chore was to clear the original shaft of debris. With the hole untouched for so many years, it took a work crew almost four weeks to arrive at the same point where the Onslow group left off.

Engineers devised a drilling method similar to that used in drilling for oil. A large drill bit augured into the hole. As it rotated, samples of oak and occasional pieces of undisclosed metals — perhaps pieces of ground coinage — were found in the auger's tailings. The engineered improvements over other methods used to date supplied added hope for positive retrieval. Possibly, the lost treasure was now within reach!

A small linked chain, later thought to be part of an officer's epaulet, appeared in the soil samples. After careful inspection, the design of the chain showed signs that it was possibly made in Spain c.1600. In addition, pieces of coconut fibers were found, indicating that a portion of the fine originated in the warm waters of the Caribbean or South Pacific. Was this another cipher that the treasure chest was near?

The flooding problem persisted. Pump hoses were immersed to siphon off the seawater. They proved only to be only partially effective. Engineers noticed that the volume of water was significantly reduced during low tides. This indicator, identified the necessity of placing a plug in the flood tunnel to gate-off the flow of water.

Countless hours of labor continued, probing for the location of the elusive tunnel. At a depth of 35 feet, near the mouth of Smith's Cove, the men thought they finally found the opening to be the source of the flooding problem! Efforts to plug the hole provided little help in reducing the water flow within the main shaft. Later investigation revealed the hole must have been part of an expanded arterial flood design tying itself into the primary tunnel — all aiding to the continued flow of seawater.

Since the tunnel plug idea failed, Truro engineers came up with a revised plan. They decided to sink another parallel shaft to avoid the flood water problem completely. Like the first, the secondary hole concept was a complete disaster. Improper shoring of the main shaft permitted the walls of the original pit to topple. Unsuspecting mud and debris filled "the money pit," destroying all evidence of possible detection.

The Truro Company, like "the pit," was eventually crushed both physically and financially. Added attempts of the company to salvage any part of the venture was considered futile by the stockholders. The entire group disbanded.

As the years rolled by, several more documented efforts were made by various individuals and organizations to find the treasure. Each group, well fortified with charts, graphs and notes from prior undertakings, guaranteed their investors the secret of success! Each try failed, ultimately resulting in bankruptcy.

During the early days of the 20th century, a notable by the name of Franklin D. Roosevelt (1882-1945), found interest in the mystique of Oak Island and what it had to offer! As a boy, Franklin and his family often vacationed in Nova Scotia. He read reports written by local newsmen that fascinated his interest in the treasure fortunes expected to be found in nearby Smith's Cove.

Roosevelt, along with three other friends bought shares of stock valued at one dollar each in the recently formed Old Gold Salvage & Wrecking Company. The men could not resist the temptation and paid several visits to the excavation site during the summer of 1909. Unfortunately, despite all the expertise generated by the principals of the company, only a mere $5,000 was raised for the research project. With the company totally underfunded, their efforts too failed like so many others.

The mystery pit took its toll of life too. Accidental drownings accounted for four deaths. One man reportedly fell, or was pushed into the pit. While working next to a steam-driven winch, the sixth man met his demise from a ruptured high-pressure steam line that scalded him so severely that he died the next morning. Were all of these bizarre episodes in defiance of uncovering the cache?

Modern-day recovery techniques are presently being used to

uncover the ongoing mystery. The borehole has exceeded 230 feet. Remote-operated TV cameras have been lowered to gain closer views of walls, oak-beamed platforms and flood tunnels. Geological tests prove samples of the recovered artifacts are definitely man-made. Archaeologists working the site, estimate that the entire project originated during the early 1600s. The eye of the camera also revealed empty chambers just below several of the oak platforms, some caused by underground erosion, some, man-made. These open voids explained why former work crews believed they had found the actual treasured vault.

Whoever masterminded the Oak Island treasure pit was a person of resourceful means. The pit obviously took months of careful planning and preparation. Hand tools of the day were limited to painstaking, laborious manual digging. The booby-trap devices found were not just accidental — traps were well thought out, planned and executed to avoid the treasure's detection.

Several personal fortunes have been spent trying to uncover the secrets of the pit. Is there a treasure chest buried beneath all of this mystery or is the entire project a hoax that was dug for the amusement of a lonely pirate? The digging continues and someday a lucky treasure hunter may yet unravel the truth!

Sultana,
A Ship That Was
A Flaming Coffin

The Civil War was over and the soldiers were coming home. But for the more than 2,400 soldiers and civilians squeezed aboard the Mississippi packet *Sultana,* the worst was yet to come. Soon 1,547 would perish in the fiery horror that consumed the near-new side-wheeler, and the hell that had been the war between brothers would blanch in the flaming tragedy that became America's worst river disaster in history!

WHEN JOHNNY COMES MARCHING HOME AGAIN HURRAH-HURRAH! WE'LL GIVE HIM A HEARTY WELCOME THEN, HURRAH-HURRAH!

The Civil War was over and the cannon at last fell silent after four long years of combat and strife. Once again the Johnny Yanks and Johnny Rebs would unite under one flag as one nation. But as the nation rejoiced in the peace after Appomattox, there still were tears and sadness, for President Abraham Lincoln was dead — the victim of an assassin's bullet on April 14, 1865.

Nor were the tragic legacies of the war soon to be banished from the headlines of the nation's newspapers, for barely two weeks prior to Lincoln's death, the troop-laden steamer *General Lyon* was burned to her waterline off Cape Hatteras en route from Wilmington to Fortress Monroe. Aboard the steamer were more

400 Union soldiers, many of them recently released Union prisoners, plus more than 100 civilians. The fire was sudden and the heavy winds soon had the ill-fated *General Lyon* enveloped in flames. Out of the 500 that were aboard that tragic night, only 20 survived the holocaust. One unit — the 56th Illinois — was nearly wiped out, loosing 11 officers and 195 men.

The nation was shocked by the loss of the *General Lyon* and so many luckless soldiers who had endured so much in combat or southern prison camps. As the cries for a complete investigation into the tragedy began to mount, the shocking news of Lincoln's assassination swept all mention of the *General Lyon* fire from the headlines.

While Lincoln's funeral train slowly crept westward, stopping at every town and hamlet to let stunned citizens pay their last respects to their fallen president, a new horror erupted in the headlines. "Thousands Dead in Riverboat Fire" blared the *Baltimore Sun* on May 1, 1865. "Calamitous Inferno on the River" headlined the *New Orleans Times*. "Fire Sweeps the Packet *Sultana*" announced the *New York Tribune*. A saddened nation already shocked by the untimely death of its elected leader was further stunned by the fiery holocaust on the Mississippi not far from the fertile soil where Lincoln was newly laid to rest.

The horror was not over emphasized, for the country soon realized that the worst disaster in riverboat history had happened just outside of Memphis. The time for national mourning was not yet past.

On the bright, clear morning of April 21, 1865, the St. Louis and New Orleans riverboat, the 1,700-ton, two-year-old *Sultana* had cleared New Orleans for Cairo, Illinois. While her desirability as a passenger side-wheeler was hawked to river travelers, she was running this trip almost exclusively for the Army.

Emaciated, weary, half-sick federal prisoners of war had assembled at Camp Fisk, on the environs of Vicksburg, Mississippi, from the pest holes of Andersonville, Macon and Castle Morgan prisons. At least 1,866 of them — $5 a head for enlisted men and $10 for officers — would pile aboard the *Sultana* at wharves that had been wrested less than two years previously from their Confederate defenders.

They came to Vicksburg in broken-down carts, their legs dangling lifelessly over the sides, on small clanking trains, lying on the rooftops, on foot, hobbling, old bloody bandages yellowing. They leaned on each other for support. Some were blind, some were armless, some were legless; all were scarecrow-thin, emaciated, starving. The great American Civil War was over, all the battles and the dead, the causes eternal, slipped into sullen silent history, and those Union troops stumbling toward the teetering docks of Vicksburg had only the acid memories of Confederate prison camps at Catawba and the cesspool that had been Andersonville. These thousands of men in tattered blue uniforms welled up like lemmings on the docks, waiting for steamboats to carry them north.

As thousands of ragged troops waited on the docks, small steamboats began to appear, and the expatriated soldiers calmly filed on board to standing-room-only spaces. The *Henry Ames* took aboard 1,300 men and paddled upstream. More than 700 soldiers crammed themselves onto the small *Olive Branch*, and that steamboat, too, moved north. The *Sultana* appeared.

Soon more than 2,000 men crossed the *Sultana's* gangway. There was also a small group of "reb" POWs on board, plus two infantry companies to guard them and protect the *Sultana* from diehard bands of guerrillas still lurking along some sections of the western shore of the river.

In addition, 12 ladies of the Christian Commission, who had served as nurses in the Vicksburg hospitals, were aboard, on their way back north. And this wasn't all! The skipper, J. Cass Mason, a part owner, found space for 236 civilians, paying full passenger fare.

In the packet's holds were 150 horses, cows, and pigs, also 150 hogsheads of sugar. The latter was partially for ballast since the steamer, like all of her sister ships on the river, was top-heavy. A huge manifest of passengers only worsened the boat's stability, since travelers has a distressing affinity for suddenly shifting from side to side in large, impulsive groups.

Overloading the *Sultana* by the transportation officer was, at least, suspect, especially since two other large riverboats had pulled to the docks to take on troops and remained empty. A board of inquiry later determined that this Union officer had been charging the owners of one steamboat line one dollar per head to put soldiers on their boats and when this was discovered, making for near mutinous cries from the sweating ranks, the nervous officer merely ordered every man remaining on the docks onto the creaking *Sultana* to dispel any idea of collusion.

Though a comparatively new boat, the vessel had trouble as soon as she departed New Orleans on April 21, with her boiler plant in need of much repair. Engineers hurriedly went to work on one boiler as men from once-proud Michigan, Indiana and Ohio regiments wordlessly crossed the gangplank to the *Sultana*.

Exactly how many troopers jammed themselves aboard the steamboat is debated to this day. Loading officers at first regulated the number boarding the *Sultana*, making accurate roles and assigning Ohio men to the hurricane deck, Michigan troops to the main deck and Indiana soldiers to the promenade deck. This soon got out of hand when the officer in charge ordered everyone on board the boat. Men then broke rank and swarmed onto the vessel in fast-moving groups, throwing planks over to the boat and crossing on their own. The pandemonium made an accurate check of passengers impossible.

As the *Sultana* cleared the Vicksburg wharf on April 25, 1865, she was low in the water, steaming toward Cincinnati. A clerk on board reported hours later to an officer that by his count there were 2,400 soldiers on board and 180 civilians, in addition to the animals and cargo. "If this vessel reached Cairo, Illinois," he said in a prosaic tone, "the *Sultana* will have made Mississippi riverboat history. It will have carried the greatest passenger load since travel on the river began. However," he prophetically added, "the vessel will never make it to Cario."

Sultana

Huddled everywhere on the open deck of the *Sultana*, the troops broke into animated conversations, songs, the first signs of spirit they have shown in the years since their misery in prison camps began. There was a plunk of banjo and a tweet of harmonicas. Rippling through the ranks ran *"Sweet Low Sweet Chariot'"* and *"When Johnny Comes Marching Home."*

Men took off their knapsacks and shared the rations they had been given in Vicksburg. Some even fed the alligator yawning in his wooden cage. Others gambled with the large amounts of back pay they had been given in Vicksburg. The river seemed quiet and rolled easy under them as the *Sultana's* paddle wheels methodically churned.

Eighteen hours after leaving Vicksburg, the steamboat berthed in Memphis. It was 7:00 p.m., April 26, and the *Sultana* had seven more hours to live. Scores of men left the boat in Memphis to visit an alley lined with 14 saloons known as "Whiskey Chute." Some of these soldiers got abysmally drunk and passed out before they could return to the boat. Years later they would thank the cheap alcohol that blotted out their minds.

The *Sultana* crossed the river to a barge and took on bunker coal while engineers continued to labor over one of her malfunctioning boilers. While taking on fuel, the steamboat regained one of her saloon-visiting passengers. A soldier appeared in a skiff rowed by a Negro. He paid the ex-slave $2 to get him back on board the vessel before she departed.

Captain W.S. Friesner of an Ohio regiment, who had watched the wharf gangs unload the ship's sugar cargo, saw the lone soldier arrive at the steamboat and climb aboard, telling his companions how "lucky" he was not to have missed her. Freisner went to his stateroom and, fully dressed, fell exhausted into his bunk.

With her coal bins full, the *Sultana* moved back out into the middle of the Mississippi. A cavalryman named Perry Summer, from Indiana, found room on the deck to sprawl, going to sleep as darkness engulfed the steamboat. It was now midnight.

Among the other "Johnny Yanks" was 21 year-old Chester

Berry, from South Creek, Pennsylvania; P.L. Horn, the same age, and Otto Barden, 24, both of Wooster, Ohio; as well as an Indiana Lieutenant, Joseph Taylor.

Their berthing was varied. Berry and Taylor had been assigned cabins, although individual space was at a premium. Taylor was not able to lay claim to a cot until he had threatened physical violence.

Barden slept atop a grating directly above one of the greasy, noisy steam pistons, but he didn't care. He'd seen some tough fighting — in fact, he had been captured by General Nathan Bedford Forrest's hard-riding cavalry. Now, any kind of accommodation looked good to him.

Horn, recent survivor of a prison train wreck, stretched his bedroll on a upper deck and dozed, oblivious to a light drizzle continuing into the first minutes of Thursday, April 27.

Those who were not asleep watched the street gaslights of Memphis fade astern in the night's damp gloom.

The *Sultana* passed the mouth of the muddy Wolf River and neared Paddy's Old Hen and Chicken Islands in a two-mile-wide stretch of the river, somewhat more than three miles north of the city. An ironclad, the USS *Essex*, at anchor just below the grubby islands, logged the sidewheeler's passage.

Just after two o'clock, when the *Sultana* was abeam the first island group, something happened! The officer of the watch on the *Essex* noticed it, shouting:

"A steamer's blown up!"

Coming downstream, headed for Memphis, was a large new sidewheeler, *Bostona.* In her wheelhouse was young Loftus Keating. Next to him stood the captain, Jules Watson. Piercing the darkness ahead, Keating suddenly saw a bright flash that turned the blue-black sky a lurid red. "That must be some man's cotton gin going up," he told Watson.

On board the *Sultana*, Captain Friesner was jarred from his sleep by what he first thought to be a riot among the soldiers. He bolted for his stateroom door, ready to reprimand the quarreling soldiers. Steam, smoke and huge tongues of fire met him on deck.

Panic screamed lustily through the ranks of the soldiers as they hurled themselves into the river. The *Sultana* was sinking, and for a fearful moment as he stood at the guard rail, Captain Freisner remembered he could not swim.

Private Summer had no such problems. He awoke to find himself already in the water, blown there by a tremendous blast. As he thrashed about, he saw the *Sultana* become a mass of flames. Everywhere about him, hundreds of men called for help.

Seconds before, the boiler in the steamboat's bowels had exploded with such velocity that it ripped apart the hurricane deck, sending splinters, steel shards and hundreds of sleeping bodies high into the air. Many fell into the river and many more, some in bloody pieces, rained back down upon the already flaming ship. Those below fared little better. The flame-belching explosion collapsed the wheelhouse, the texas, and one of the smokestacks. This weighty blazing debris showered down on hundreds of sleeping troops, scalding and burning them to death.

The fire did the rest. Flames, beginning about one-third of the distance from the bow, were driven toward the stern by downstream winds. Fires ate their way to the staterooms and promenade deck, forcing soldiers to jump. Many of these were amputees who had no hope of surviving in the water. Throughout the ship one vast horror unfolded as men attempted to save themselves.

Otto Barden, close as he was to the engine room, survived what apparently had been a tremendous blast of the leaky boiler that had been patched up in Vicksburg. Some of the details, as he recalled them, follow:

"Hot steam, smoke pieces of brickbats, and chunks of coal came thick and fast. I gasped for breath. A fire broke out that lighted up the whole river. I stood at this hatch-hole to keep comrades from falling in, for the top was blown off by the explosion. I stood there until the fire compelled me to leave. I helped several out of this place. I tried to get a large plank, but this was too heavy, so I left it and got a small board and started to the wheel to jump into the water. Here a young man said to me. 'You jump first, I cannot swim.'

"This man had all of his clothes on. I had just my shirt and pants on. I said to him, 'You must paddle your own canoe. I can't help you.' Then I jumped and stuck to my board."

Berry was struck on the head by a piece of cord wood. He was sure it had fractured his skull, noting:

"The first thought I had was that while the boat lay in Memphis someone had gone up the river and prepared such a reception for us. I lay low for a moment, when the hot water soaking thorough my blanket made me think I had better move, I sprang to the bow of the boat, and turning, looked back upon one of the most terrible scenes I ever beheld.

The upper decks of the boat were a complete wreck and the dry casings of the cabins, falling in upon a hot bed of coal, were burning like tinder. A few pails of water would have put the fire out, but alas, it was 10 feet to the water and there was no rope to draw with; consequently the flames swept fiercely through the light wood of the upper decks."

He might have added that all of the wooden buckets were themselves burning like torches.

Lieutenant Joseph Taylor noticed that many men remained asleep after the explosion. The account of his experience is graphic and horrifying:

"The thought came to me that I had a nightmare, and in that condition of mind I turned around and made for the stern of the boat, hardly knowing what I was doing. The ladies' cabin was shut off from the men's cabin only by curtain, and I pushed back a curtain and pushed through, when I was confronted by a lady, who I suppose was in charge of the cabin, with: 'What do you want in here, sir?'

"I paid no attention to her, but went ahead, saying that there was something wrong with the boat. I went on through the cabin to the stern of the boat and climbed up to the hurricane deck. Throwing myself across the bulwark around the deck, I looked forward toward the jackstaff. Such screams I never herd — 20 or 30 men jumping off at a time — many landing on those already in the water

— until the river became black with men, their heads bobbing up like corks, and many disappearing never to be seen again. We threw over everything that would float that we could get hold of, for their assistance...

"Soldiers crawled wildly over their companions, friends who had been burned and scalded and pleaded for help. Screams of agony surmounted the roar of the flames. 'I couldn't stand it said one man later, 'the stench of burning flesh was intolerable.' In mad scrambles men raced from stern to bow, and back again, always met by the flames and trampling others underfoot. 'They jumped overboard as fast as they could,' one report states, 'tumbling into the river upon each other and going down into the deep by the hundreds.'"

Captain Friesner was one of these, but he took the precaution of ripping off his stateroom door and diving into the river with the door tucked beneath him. He used the door as a raft and paddled away from the mobs of drowning men who were floundering and taking strong swimmers to the bottom in death grasps.

The *Bostona* appeared in the distance. At her wheel Keating pointed out the wildly burning *Sultana*. "Why, just look at the cattle jumping into the water," he shouted to Captain Watson.

Watson leaned forward, peering. "My God, them ain't cattle. They're people! Holler down to the engineer. We must get there as soon as we can. Why, just look at them jumping!"

Private Summer saw the *Bostona* coming downstream, but by then he had already secured two planks, and holding one between his legs and the other under his arms, he paddled to shore. All about him men were coughing and dying. He saw a horse swim by with 10 men trying to hold on to it. They went under en masse. A friend sitting on a barrel and using his feet as oars, splashed by him. The man's face had been burnt away; his eyes were gone. Summer drifted with the current for an hour, sliding with the river back to Memphis, where he was plucked from the river by a Negro. Once in the skiff, Summer realized that almost every rib in his body was cracked.

A large slab of the boiler deck, which had been sheared away from the *Sultana* in the blast, was still intact. A dozen men, who had been sleeping on it, used it as a raft to sail to Memphis; none were injured. For an hour, close to 700 soldiers took refuge on the steamboat's bow as the fire consumed the aft portion of the *Sultana*. The wind then shifted, and the blaze swept toward them. In one agonized rush, it swept the men from the bow of the boat and into the water where hundreds perished.

The *Bostona* bore down on the flaming wreck, its fires lighting up the river for miles and illuminating hundreds of bobbing heads in the water. Watson ordered the yawl lowered, and scores of men were soon picked up. Captain Friesner simply paddled his door to the side of the steamboat and was dragged upward to her cargo deck by a roustabout using a gaffhook.

The sidewheeler burned for an hour and a half. When survivors began drifting into Memphis, the *Essex* and *Tyler*, along with a half-dozen smaller boats, got up steam and put into the river in search of more soldiers. More than 600 persons were brought into Memphis, but almost 200, the flesh of their bodies burned away, died in makeshift hospitals. For days steamboats and barges edged along the riverbanks, finding bodies wedged in thickets and plied on sandbars. All of the public buildings in Memphis were opened and used as charnel houses.

With a "hissing," the nearly incandescent *Sultana* vanished beneath the river. Since her paddles had stopped with the explosion, she had floated like a derelict almost two miles back down stream. In moments, only the flaming fagots of her superstructure cast their ghastly, flickering shadows over the horror still being enacted upon the predawn surface of the "Father of Waters."

Otto Barden, who had clung to a tree, was rescued by the same steamer, as was Horn.

Daniel McLeod, too, was found in a tree, half submerged on the river bank. (He made a full recovery after the amputation of his right leg.)

Nate Wintringer, the chief engineer, had saved himself by hanging on to a plank. A little girl, barely clinging to another flotsam but splashing with her free hand, vanished beneath the murky waters as a boatman vainly grabbed at her.

Only one woman was known to have survived, although there were rumors that one other had come ashore at the foot of Beale Street, after having lost hold of her child. The deliverance of the sole identified lady was unusual. Mrs. Harvey Ennis, who had boarded at Memphis, rode in plunging fashion to shore on the broad back of a mule. Her husband, a lieutenant in the federal Navy, her child, and a sister all drowned.

The *Sultana* had come to rest. The charred hulk settled in a depth of 26 feet, her jackstaff showing. At low water, bones, some skulls, a scorched shoe, and blackened rifle muzzle were visible within the wreck. None came to visit the gruesome spectacle.

Bodies continued to wash ashore for weeks down the snaking course of the Mississippi to New Orleans. The body of the captain was never located, and thus a $200 reward for J. Cass Mason was never claimed.

The final count of the dead remained long in debate. An Army board of review sought to minimize the disaster and claimed that only 1,238 men had been killed. This figure was refuted, however, by Memphis courts holding open hearings; the death toll was placed much higher. Officially, the United States Customs Service at Memphis finally released the total death count: 1,547. This number represents to date one of the worst marine disasters in the world and exceeds by 30 the *Titanic*.

The totals could have been even higher. It was, by any count, the worst river disaster in American history — an unenviable record that is yet to be exceeded.

Several years later a diehard Confederate in Vicksburg boasted he had placed a bomb in the *Sultana's* coal bins. He was not taken seriously. An imperfect patch on a leaky boiler, enabling it to explode under pressure, made more sense to the river men who sorted the grim evidence.

For years after the disaster, members of the *Sultana* Survivors'

Society met annually, members reminiscing over the disaster and marveling at their own escapes. At one session someone remembered the alligator that had been on board. Everyone felt sure that the beast had certainly survived. Then an Ohio man stepped forward and sheepishly confessed that the alligator died with the *Sultana*. Of this he was positive. He admitted sorrowfully that during the raging fire, he became a beast himself, stabbing the alligator to death, taking its wooden box and using it to float to safely.

Oddly enough, almost 125 years later modern technology has yet not discovered the charred skeleton of the *Sultana*. Although the remains of this doomed vessel are said to lie in the mud across from the modern Memphis Civic Center, numerous attempts to locate the wreck and its supposed $20,000 in gold have ended in failure. Farther upstream, well beyond the shallow inlet known as Hopefield Bar, pilings sunk for a new bridge in 1970 encountered no evidence of a rotting steamer's rusty carcass. Wherever the remains of the *Sultana* lay in the muddy depths of the Mississippi they are ably protected by the ghosts of the men who never made it home. Their forlorn distinction lies in the record book as being the hapless victims of the worst river disaster in American history.

The Ghost of *Bannockburn*

On the morning of November 21, 1902, Captain George Wood gave the order to castoff the dock lines of his Great Lakes grain steamer, *Bannockburn*. The weather on Canada's northeastern side of Lake Superior was the typical ominous gloomy overcast with light rain showers. There was talk of the possibility of light snow by nightfall.

Once the ship cleared the dock at Midland, Ontario, the order was given to head into the lake at slow speed. Entering deeper waters, Captain Wood found the *Bannockburn* beating into high seas with winds approaching near gale force.

Standing his watch on the bridge, Wood looked out as his deck crew scurried about lashing last-minute items to cleats. Although the crew disliked the bitter wind chill, this was the fall of the year — Lake Superior was known for its fearsome weather, but this is what the men were paid for.

The *Bannockburn*, a lake grain steamer, was operated by a single compound coal-burning steam engine. She measured 244 feet in length, with a 40-foot beam, displacing 1,650 tons. The vessel built in Scotland in 1893 was owned by the Montreal Transport Company. The owner was known for its ships that moved assorted grain cargo over the Great Lakes.

Her position was reported as 50-miles south-southeast of Passage Island en route to Ft. William where she was to unload her full load of grain.

With seas still running high, it was late afternoon of November 21, 1902, when the crew of *Algonquin,* a similar steamer, sighted the *Bannockburn* sailing in the distance. The two ships followed parallel routes, often passing each other as they worked the ports of Lake Superior.

Despite the adverse weather of continued high seas and gale winds, her gray silhouette was identified as the *Bannockburn* by James McNaught, captain of the slow-passing *Algonquin*. He noted the sighting in his log and that she appeared to be running low, not towing her usual barges.

McNaught reported that he turned his back to check their position on the chart. No more than a minute or so later, the gray silhouette vanished from his view! With field glasses in hand, he scanned the horizon several times carefully rechecking — there was no question the ship was gone!

He recalled saying to his mate, "She's gone! I can't believe this."

The mate looked at McNaught. "What are you talking about, captain? What's gone?"

"That ship out there — she just vanished! I'm sure it was the *Bannockburn*."

Captain McNaught gave the order to turn toward the last area of sighting. He thought it strange the ship disappeared in such a short period of time! However, it was possible the *Bannockburn* slid behind a fog bank. His mind questioned, however, heavy fog was normally associated with becalmed waters and not on blustery seas such as they experienced this afternoon. McNaught ordered the engine room to operate at half speed fearing the *Bannockburn* might suddenly burst into view.

It was now late in the day, visibility remained rather poor, there was no sign of fog as they searched. It was possible to see a faint gray outline of the northerly shoreline. Much to his regret, Captain McNaught felt the *Bannockburn* sunk before him without a trace.

During the turn of the century, most ships operated without the aid of wireless radios. There was no way of making formal contact with other vessels or with the home port to confirm the disappearance until they actually arrived dockside.

Upon reaching port, it was confirmed the *Bannockburn* was

overdue and feared lost. A formal search was already set into motion.

Two company-owned tugs, the *Boynton* and *Favorite,* were dispatched, initiating an immediate search in the area of her last reported sighting. As they scoured the sea, no traces were uncovered. It was a complete mystery as to her disappearance. No reporting or any form of a distress signal was seen by other ships or land observers.

Confirming *Bannockburn* was lost at sea, several conflicting reports were received. One, she ran aground along the north shore of the lake. A steamer, the *Germanic*, said she was swamped by the large storm and rested disabled north of Michipicoten Island. Hearing this news, a rescue party was immediately formed — nothing was found! This made no sense to Captain McNaught; if a ship did run aground, surely he would have seen it from his vantage point.

Still another conflicting sighting from shore observers came — the vessel went aground near State Island, in Whitefish Bay. After checking, it too proved to be untrue.

The only report of unidentified flotsam was found by the captain of another passerby lake steamer, the *Frank Rockefeller.* His crew found remnants of what appeared to be parts of a ship's superstructure, but they had no idea as to when or where it originated. Their only observation — it was in the vicinity of where the *Bannockburn* supposedly went down.

The *Bannockburn* simply vanished! The weather started to improve and still there were no visual sighting or reports from the newly established coastal wireless radio stations within the area.

One theory held that the high seas were the cause of the *Bannockburn* to run off course. Wood may have feared the heavy load of grain in her hold would cause a problem due to the storm. Visibility may have been reduced to a point where she ran a reef off Caribou Island and sank. Still speculating, Captain Wood nor any member of his crew failed to see the Caribou Light — it was inoperative during the time of her voyage crossing the lake. Wood may not have known about the outage.

Another possibility was that the *Bannockburn* dropped her boiler and engine machinery through the thinning hull of the ship. It may all have happened at one time accounting for the quick disappearance of the vessel. Although the ship was still considered relatively new and holding a high Lloyd's of London rating, a large exterior hull plate was later found while servicing the lock at Canadian Soo. Workers later said, this portion of the hull was identified as coming from the *Bannockburn*.

Also, prior to departing Midland harbor, the *Bannockburn* ran aground and was stranded for more than an hour. Captain Wood jockeyed the ship's engine back and forth trying desperately to free her of the grounding. During this maneuver, portions of the hull may have been stripped from the bottom, unbeknown to the captain or crew. It was reasonable to believe several portions of the hull were loosened or lost, therefore and when the ship hit the reef the entire power unit may have fallen through the hull. With its heavy cargo of grain, the ship could easily have sunk in minutes and remained totally undetected.

Captain McNaught dismissed this idea since the ship did vanish quickly. He also concluded that if her boilers had exploded from the onrush of seawater, surely, he would have seen some trace of smoke? It all happened too fast.

Finally, several weeks after the disaster, a fisherman found a single life jacket marked *The Bannockburn*. This was the only remnant uncovered.

To this day, seamen standing watch while running along the north shore of Lake Superior report seeing the ghost-like image of a vessel resembling the the *Bannockburn*. It too remains for only a short period of time before it disappears into the night sea.

The Fate Of The *General Slocum*

It was June 15, 1904, a beautiful summer morning when nearly 1,400 men, women and children boarded the 250-foot excursion side-wheeler, The *General Slocum*. Most of the passengers were of German decent, stemming from neighboring Lutheran Churches. This was considered the outing of the year when friends and families could join together for a day of festivity on the river. Church pastors and their families acted as the greeting committee while fellow parishioners boarded the ship. Everyone offered smiles, hugs and kisses showing their pleasure.

This was the day of the annual picnic for Long Island's Locust Grove Church, which in collaboration with St. Mark's Lutheran Church, made the outing possible.

On deck, a band played in celebration of the happy event. The conductor, standing in a colorful uniform gayfully directed songs from German marches as passengers found their way to deck chairs. Children ran about dodging fellow passengers carrying baskets of food and sunshades. Little English was spoken; songs were in German, small groups gathered dancing to folk music or joined in barbershop-like fashion.

Captain Van Schaich stood in command at the wheelhouse. Satisfied all passengers were finally aboard, he looked at his pocket watch; it was almost 10 a.m. He gave the order to push off from the Third Street pier and head his day-liner up the East River from Manhattan's lower East Side.

Clear of river traffic, the *Slocum* backed itself slowly into the

East River. Van Schaich then ordered engines all ahead into the river's calm water. The large side wheels stopped, reversed direction and throbbed forward. The ship, owned by the Knickerbocker Steamship Company, planned to sail up the East River, through Hellgate and into the Long Island Sound, where she would make a turnabout, then return to the pier before sundown. Captain Van Schaich, a seasoned officer with years of seafaring experience, mastered the river day-liner. He considered the excursion an easy task for his crew. With arms folded, he stood proud commanding his vessel. Looking on deck, he waved from time to time as small children played before him.

The ship, built in 1891, was named after the famed Civil War hero, General Slocum. Her power was generated by two compound steam engines, driving state-of-the-art side wheels. She displaced 2,200 tons. Her all-wood construction provided three open observation decks offering panoramic views for her 2,000 guests. Knickerbocker Steamship considered the *Slocum* the pride of their fleet boasting a safety record of carrying over two-million passengers!

New York's East River had its usual load of river traffic — tugs toting barges, cargo-carrying steamers and pleasure boats making way. Van Schaich pointed to several river vessels providing added warning to his helmsman.

Now traveling at a respectable 11 knots, the *General Slocum* was only five miles up river from her point of departure when a passerby tug noticed smoke stemming from the lower center deck of the *Slocum*. The tug's captain knew immediately something was wrong! He gave three short blasts on his ship's whistle, signaling the distress signal.

Hearing the blast of the horn, Captain Van Schaich turned about only to find smoke pouring out of the stack on the center deck. Realizing there was an immediate problem, he too sounded the ship's alarm with a continued three shorts blasts of the ship's whistle. Now, every ship in the vicinity was aware of the *General Slocum's* perilous condition.

General Slocum

One parent recalled her little boy tugging at her dress pointing, "Look momma! There's smoke and fire over there." Within seconds, flames broke through the wood decks of the *Slocum*. Passengers were horrified running about confused as to what to do and where to go! Children and old folks alike, clung to family members.

Below in the ship's galley, one of the hands found smoke filling his galley station as he worked. In his haste to run, he accidentally knocked over a lantern filled with kerosene, fueling still another fire from the hot stove. He began shouting, "Fire, fire!" Now, passengers became frantic with the fearsome call.

On deck, leaders of the church cried out to their parishioners to be clam, not to panic. Something would be done quickly! Their chaotic words were no sooner spoken when flames roared through the wood decks. A crew member holding a fire hose yelled, "Put on life jackets — jump into the water!" The fire hose he held proved useless, as most of it had burned before being used.

The flames fanned by the ship's forward 11-knot speed, quickly sending itself to the long dresses worn by the women and children. Having little other choice, flaming passengers began jumping into the water for fear of burning alive. Those who used life jackets found they were of little use. Their cork inner lining formed a dry-rot condition, disintegrating immediately upon hitting the water. The wet jackets added to the individuals' weight pulling many under immediately. Most of the floundering passengers were unable to swim because they were too young, too old or too weak. Those strong enough to survive were pulled down by others less fortunate groping for their lives. Bodies were seen everywhere. Hot embers from parts of the ship fell into the water, scalding several individuals.

Wolfgang Mueller, a survivor, recalled seeing his "frau" (wife) jumping into the water before him with their son and daughter in hand. The three perished before him in the swirling water. The man was helpless; he too could not swim!

With the fire now reaching the base of the wheelhouse, Van

Schaich ordered his helmsman to turn toward shore. Beaching the ship in shallow water in hope of reaching a nearby dock. He believed his passengers would have a chance to make it to safety. A longshoreman standing at a nearby quay held a horn in hand calling to avoid docking at all cost. "The dock is filled with barrels of oil. Don't dock here!" He shouted, waving his arms frantically hoping to attract the attention of the captain. Taking heed to the man's plea, the helmsman made his turn toward the beach completely avoiding the dock.

The ship became a floating inferno. Now close to shore, the heat was so intense it was near impossible to offer assistance to those pleading for help.

The tiny steam-powered tug, *Walter Tracy*, took the initiative to come alongside to capture as many of the screaming passengers as possible. With her deck filled with passengers, many of whom were badly burnt, the captain of the *Tracy* was forced to back off fearing more survivors might swamp his small boat.

In her futile attempt to help, another small ship, the *Goldenrod*, pulled alongside the burning hull. She too feared being caught by the flames, but the captain thought his ship was the only hope for many who stood in desperation of their lives. Several jumped to the deck of the *Goldenrod*, many braking limbs in their fall. Crewmen did their best to catch and break the fall of others. Only minutes passed when the *Goldenrod* too was forced to back away due to the intensity of the flames. A chill ran down the spine of the captain knowing there little more he could do. Others still clung to their last hope of life screamed for help while remaining on deck of the *Slocum*.

Now, as shoreline onlookers watched the spectacular display, several people waded into the river to aid in the rescue of passengers. Most of the bodies claimed were charred from the flames. One man reported, "Charred body parts were found everywhere. It was a sickening sight to behold! I've never seen so many torn arms and legs before."

In another accounting, two prisoners from nearby Rikers Island prison risked severe punishment from their captors as they ran off, grabbing a small rowboat to offer assistance. The prison tower guard took aim with his rifle, but held his fire as the men began hauling floating bodies into their boat and rowing them back to shore. He watched as the task repeated itself several times over. The two were later cited for bravery and were given a reduced sentence for their heroic act.

One man, Heindrich Kohler, an insurance investigator, perished in the flames along with his wife, son, in-laws and other relatives. In all, 29 members of the Kohler family died, the greatest number of a family to perish in a single marine tragedy.

By midnight it was over. All that remained was the smoldering ashes of a one-time glamorous river excursion boat. The fun, thrills and excitement of the day came to a horrible end. Some 1,030 passengers perished by fire and/or drowning, another 175 severely were burned or injured from various mishaps. Of the balance of survivors, still another 77 died later from various injuries caused by the disaster. It took weeks for police to identify the remaining bodies and several passengers were never accounted for.

To the surprise of many, Captain Van Schaich survived the holocaust. Although badly burned, he did manage to dive overboard, claiming he was that last of the crew to leave the ship. While still in the hospital, the outraged American public screamed for justice. Many onlookers felt Captain Van Schaich was at fault; it was he who failed to properly inform his crew on fire safety while aboard ship prior to departure.

A year and a half passed before Van Schaich was finally summoned to trial. The prosecution claimed he was guilty of *criminal negligence.* Standing as witness to his own defense, Captain Van Schaich swore he personally did everything humanly possibly to save the lives of his passengers and ship while under the siege of the fire. With tears rolling down his cheeks, he also stated, "I was the last man to leave the ship. This was the reason for burns over most of my body. No man could have done more!"

The jury listened attentively, but failed to see why decaying life jackets were still in use and why fire hoses were inoperative! Wasn't this someone's responsibility to see all of this equipment was in working order? Surely, it was the responsibility of the captain. Was there really a man on a wharf with a megaphone warning them of oil drums and another potential fire disaster? Would that wharf have been the salvation to many of his passengers?

Jurors left the court to deliberate their findings. Only hours passed when they returned with a verdict. Captain Van Schaich stood facing his accusers.

Their spokesman read, "We find Captain Van Schaich guilty as charged of criminal negligence. We recommend 10 years of incarceration."

Devastated by their findings, Van Schaich could not believe his ears. He waited while his court guards escorted him away.

The American public was mortified by the outcome of the trial. Most felt the captain became the scapegoat for his employer, the Knickerbocker Steamship Company. It was they who were the actual scoundrels who failed to make the *General Slocum* a safe vessel. It was they who could afford to pay attorneys to hid any fact that might place blame on the company.

Finally, in 1911, 250,000 people signed a petition protesting the unjust prosecution of Van Schaich and presented it to President, William Howard Taft. After reviewing the facts, the President agreed to executive clemency. Captain Van Schaich was released from custody after spending almost seven years of his sentence. Although he was happy of the ultimate outcome, his life did come to an end in 1927. Friends thought he died from the trauma of being unjustifiably incarcerated for something he tried so hard to avoid.

In conclusion, the actual cause of the fire was never fully determined. With the ship finally burning to the waterline, there was so little evidence left of the hull that fire officials could not positively identify exactly where or how the original fire started. Investigators commented, that while they knew the secondary fire in the galley was caused by the upset of the kerosene lantern, the main blaze was still a mystery. When the question was asked of Captain Van

Schaich, he too was unsure of the exact location or cause. He assumed the problem stemmed from or near the engine room and spread quickly. Those who may have known the exact cause were either dead from drowning or burned in the fire.

The holocaust of the *General Slocum* was the worst civilian marine disaster America had ever seen exceeded only by the burning of the Mississippi riverboat *Sultana,* on May 1, 1865, when 1,547 Yankee Civil War soldiers lost their lives returning from prison camps in the South.

To this day, the story of the burning of the *General Slocum* remains on the lips of surviving families. The mystery has been passed down through generations as many had grandparents or relatives attending that festive day.

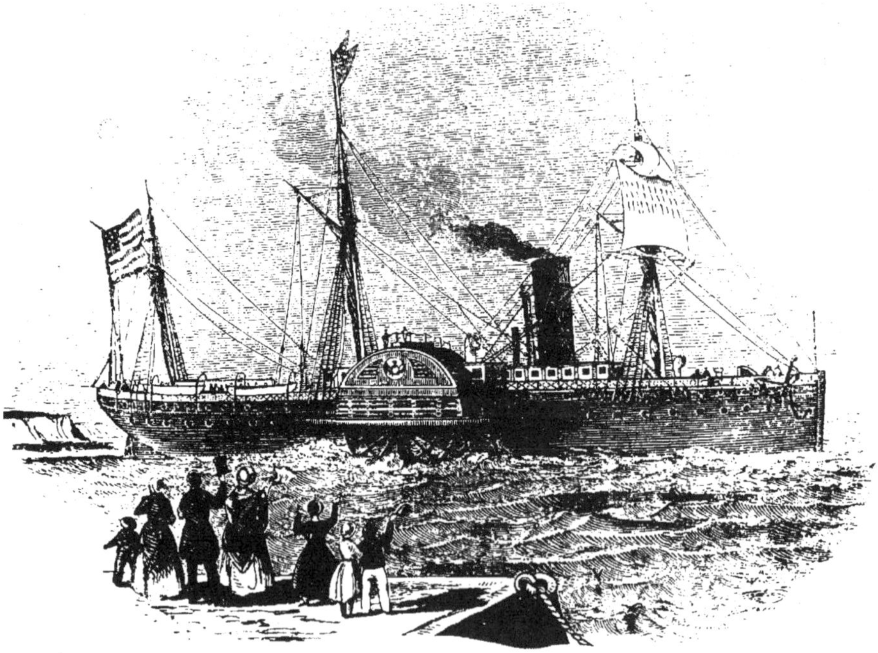

The Elusive
City Of Rio de Janeiro

It was February 21, 1901, when the liner, *City of Rio de Janeiro*, slipped beneath the water just off the Golden Gate entrance to San Francisco Bay. The 350-foot, 3,500-ton ship was returning from a long voyage from Honolulu, Hawaii, with 213 passengers and crew. Pacific Mail Line of San Francisco claimed ownership of the aging steamer. The following gives an account of what took place that dreadful morning.

Prior to the devastating event, the Rio sat at anchor off the California coast patiently waiting for a dense fog blanketing San Francisco to lift before entering the harbor. The Golden Gate entrance was known to be hazardous while fog clung to the bay. The captain decided not to risk the chance of running aground under these poor conditions. With the Rio securely anchored to the bottom just off Point Lobos, Captain William Wood retired to his cabin for the night.

At four in the morning, eight raps were heard from the ship's clapper. First Officer John C. Johnson called Captain Wood on the voice tube. "Captain, the fog has lifted, sir. Our harbor pilot, Fredrick Jordan, is on board. With your permission, I'm preparing to weigh anchor and enter port."

Captain Wood replied, "Very well, Johnson, I'll be up shortly." He dressed, then climbed the ladder to the wheelhouse. The galley steward, an Oriental man, awaited with hot coffee.

Peering through the window of the wheelhouse, lights of the city were visible in a distance. Just below, a hand wave and lantern signal from a deck member confirmed the anchor was raised and properly stowed. Wood tugged on the ship's whistle to signal they were finally getting underway. He informed the engine room to fire the boilers.

Below decks, passengers were awakened by the early morning belching of the ship's horn and then the soft pounding from the engine room confirming the ship was making its final entry to port. Several guests began to dress, looking for their morning cup of coffee and the last meal of the voyage before disembarking the ship. Those who preferred the all-night poker game yawned, making it their last hand before returning to their staterooms. With luck, the Rio would be in port well before noon.

In the engine room, chief engineer Philip Hurtihy opened the valve providing more power to the *Rio*'s engine then gave instruction to his black gang, mostly Chinese, to add coal to the boilers — steam pressure was getting low. He called over the sound of the engine roar, "It will only be a few more hours, men, and we can call it a day! Now put your backs into it!"

The Point Lobos lookout station, located at the south entrance to the Golden Gate, was used to inform the Merchant's Exchange in the city of incoming ships. Attendant John Hyslop took field glasses in hand confirming the *Rio de Janeiro* was underway. He made a call advising the Immigration and Custom's Officials of the ship's arrival — approximating another three hours. He noted in his log the ship's sighting at 4:42 a.m. Hyslop also commented that although the heavy blanket of fog lifted slightly, it was again impossible to see the Point Diablo light located directly across the inlet.

Captain Wood gazed ahead as harbor pilot Jordan provided the helmsman with directions to enter port. The *City of Rio de Janeiro* moved forward at seven knots. Sea and wind conditions were calm. The words no sooner left his lips when the ship entered another diminishing fog. Visibility fell again, this time to less than 50 yards. The concerned Captain Wood turned toward the pilot asking, "Would you like us to stop and take soundings, sir?"

Jordan waved his hand denying the offer, "That will not be necessary, captain, I know every inch of the way. Steady she goes, helmsman. We're doing just fine."

Still concerned with the limited visibility, Wood ordered the ship's horn blown in repeated intervals with hopes of warning other vessels that might have suddenly found themselves in the same situation. Wood recalled entering this port many times before and even under good conditions knew of the treacherous water surrounding the shoreline. Numerous ships went aground due to the fierce ebb tidal conditions. Wood looked, about showing deep concern for what he considered a dangerous situation.

Without warning, the ship vibrated slightly and a low groaning sound was heard throughout the ship. "What was that?" asked Wood, running from one side of the wheelhouse to the other.

"Nothing to be concerned about, captain! That was only the scraping for the Fort Point buoy along the hull," assured the harbor pilot.

Wood cried out, "Nonsense!" He signaled the engine room to make an emergency stop and take an immediate sounding off the bow.

"I think we just ran aground, mister! That was more than the scrape of a buoy."

The entire ship then shuddered, vibrating so violently that passengers and crew were knocked off their feet. Items on shelves fell to the floor, furniture toppled, glassware flew off tables and hot coffee scalded the laps of those having their early morning breakfast. The propeller strained itself turning in the opposite direction trying to halt the vessel. There was another loud grinding sound!

Rushing to the voice pipe connecting the wheelhouse to the engine room, Captain Wood called out, "Hurtihy! What the hell is going on down there?"

He waited as seconds passed. Finally the chief engineer replied, "Captain, we're hurt bad! There's a big gash in the hull. We're flooding sir! There is no way the pumps will handle the water." In the background Wood heard the screams of the engine crew as they fought in desperation to plug the cascade of water entering the ship. Then he overheard Hurtihy yell, "Alright, all you men get the hell out of here. There's noth'n we can do now! She's going down…"

Realizing the severity of his words, Captain Wood gave the desperate command, "Damn it! Abandon ship. Get everyone off as quickly as possible, Johnson."

Hearing the order, first mate Johnson ran from his post, passing the word to other crew members to warn passengers of the situation. Stewards and deck crewmen frantically ran from stateroom to stateroom pounding on doors, yelling, "Get out! Abandon ship, abandon ship! The ship is sinking."

Remaining in the wheelhouse, Captain Wood began tugging the lanyard of the ship's horn. The loud distress signal would confirm to all the serious nature of their situation. He hoped there would be sufficient time to evacuate passengers and crew in a safe fashion. Only minutes passed when his helmsman noticed the bow being low in the water. "Sir, look at the bow now!"

The gash in the hull must have been greater than Philip Hurtihy realized. The ship was going down at a rapid rate.

Turning again to the voice pipe, Captain Wood called down to his engineer, Philip Hurtihy. "How does it look now? Any more to report, Hurtihy?" He asked waiting for a reply. Wood could hear water rushing into the bowels of the ship, the sounds of steam hissing in the background and men still screaming.

Then the reply came, "Captain, so far I've lost seven of my men. We just lost the lights; it's totally dark down here. We're trying to get out!"

At Point Lobos lookout station, John Hyslop, who remained at his post, could hear the distress blast of the *Rio*'s horn. He stood with binoculars in hand trying to see anything in the foggy blackened mist. There was no visual sighting. Hyslop made another call to the Immigration Office explaining there was something wrong. A ship, which he assumed was the *Rio*, was sounding a continuing distress signal. He could not, however, positively identify the vessel in the early morning fog.

Knowing many of the passengers were still in their rooms, a Chinese steward who barely spoke English he did manage to continue yelling, "Ship sink! Ship sink!" Holding a lantern in hand, while passengers exited their staterooms rubbing their eyes, he pointed to the water now filling the halls of companionways. Panic stricken passengers, many of the still in their nightclothes, ran about in confusion not knowing which way to run.

Adding to the mayhem, the *Rio* began listing heavily to starboard as the hole in her side enlarged. The steep angle bound many of the cabin doors and hatch exits in the closed position. Between the heavy surge of water and now binding latches, doors were impossible to open. Passengers screamed for help, pounding on doors and walls while trapped inside. Foolishly, others opted to return to their cabin with hope of retrieving personal valuables. Their stupidity cost them their lives. The more fortunate passengers and crew who eventually made it to the upper deck were unable to grab hold, slipping into the icy blackened sea.

The ship's final blow was the belching roar of the boilers as they too were sucked into the water. One final explosion, a spume of smoke, then it was over — the *City of Rio de Janeiro* took her last breath of life and sunk, 5:30 that morning in total darkness.

Survivors who did manage to escape the wrath of death clung to any floating debris available. It made little difference: wood tables, deck chairs, cushions or pieces of wood torn from the ship all acted as lifesaving devices. They floundered in the stygian water not knowing which way to turn. Voices called in the dark crying for help.

In their scurrying attempt to save as many lives as possible, crewmen did manage to deploy three lifeboats from the *Rio's* midship. The last, still tied to her davits and falls, was pulled under as the ship rolled still farther dragging the tiny boat with crew and passengers clinging to her gunnels.

On the remaining two boats, oarsmen pulled, as they searched for those pleading for help in the blackened water as others dragged numbed survivors aboard. Still engulfed in total darkness, passengers could not tell the direction of the shore. It was possible that the stronger swimmers, thinking they were near the shoreline, may have actually swam toward the ocean and were never seen again!

Both lifeboats, overloaded with passengers and crew, did manage to row to safety. Exhausted individuals crawled out of the boats to rest on rocks.

Hearing the distress signal of the *Rio*, the nearby Life Saving Station sounded the alarm and launched boats in an attempt to offer aid. By the time their boats reached the scene, it was too late; survivors were either in the *Rio*'s lifeboats or had managed to swim to shore. All that remained were the bobbing dead bodies found among the flotsam.

In less than 10 minutes, the *City of Rio de Janeiro* claimed the lives of 131 out of her 213 persons aboard. Both the captain and his first mate went down with their ship. Survivors found it difficult to explain why or how the accident occurred. Many confessed that the horrifying experience happened so fast there was no time to think!

Harbor pilot Fredrick Jordan, the only man to get off the bridge alive, had serious explaining to do before a board of inquiry. It was his sole responsibility to guide the vessel safely into the harbor. When questioned why the ship was so far off course, he stated he tried to determine their exact position, but due to the quick formation of fog and then total darkness, it was impossible to pinpoint their exact location. He also gave an accounting of the verbal conversation between the captain, the wheelhouse crew and engineer.

According to his vague recollection, he calculated the *Rio* was at least a half mile or more from the shoreline when she went down. His testimony proved inconclusive and often contradictory while fumbled for words. Jordan placed all of the blame on Captain Wood who failed to adhere to his advice. Since there were no other survivors of the wheelhouse, no one could refute his statement. During the inquest, attending survivors listened to Fredrick Jordan's testimony, but they were unable to offer comment.

Confused by the outcome of the inquest, the owners decided to take action and search for the ship's remains. A salvage team was hired to find the ship and, hopefully, the ship's log and safe would be retrieved.

Divers started their search where Fredrick Jordan claimed the *Rio* had sunk. After two weeks of diving into depths exceeding 50 or more fathoms, no signs of the ship were recovered. One diver commented, "It was as though the ship had vanished completely!" They could offer no explanation.

San Francisco newspapers received anonymous information that the *Rio* was secretly carrying a large cache of gold and silver bars taken on in Hong Kong. The total value, supposedly exceeded several hundred thousand dollars! The ship was actually a treasure ship of sort! So called treasure hunters of the day became excited with the news and made dives of their own. After rechecking all possible manifests at her ports of origin, it was later determined the claim was no more than a hoax. The reason the ship's owners wanted the logs and safe was to obtain more information supplied in the captain's log and retrieve the purser's safe so that valuable items belonging to passengers would be returned to their rightful heirs.

Years after the tragedy, several marked pieces of the ship were identified as belonging to the *City of Rio de Janeiro*. They surfaced finally floating onto a beach almost 40 miles to the south of the accident. How was this possible? Was it true? Or, did someone purposely plant the items to throw off other treasure hunters?

There was much speculation as to the disappearance of the ship. The waterway entering the Golden Gate is often regarded as deep and extremely fast flowing. Many doubt if any ship would simply remain in its path. More likely, the flowing current and tides of the harbor entrance swept the *Rio* out to sea where she rests in a deep offshore grave. The elusive *City of Rio de Janeiro* will remain a mystery forever.

A typical turn of the century dayliner.

Disappearance Of The Ill-Fated *Portland*

Large posters strategically placed at railroad stations throughout the state of Maine read: *Portland Steamship Co. Offers $1.00 Fare For Passage On The* Bay State *Or* Portland *Steamship.*

Who could resist such temptation to take an eight-hour boat trip from Boston, Massachusetts, to Portland, Maine, for the meager sum of one dollar?

Built in 1890 by New England Ship Building, the *Portland* was of all-wood construction. She measured 291 feet in length with a beam of 42 feet. A single steam-powered engine developing 1,200 horsepower drove two, large-side wheels pushing her 2,300-ton displacement. Normal speed was estimated at 10 knots.

It was the early evening of November 27, 1898, when the side-wheel steamer *Portland* left India Wharf, in Boston harbor. In command, was her master, Captain Blanchard, with 176 passengers and crew. The ship, making its usual 100-mile trip north, was considered a safe, reliable link between the two cities.

Captain Blanchard checked the time confirming the 7 p.m. departure. Dock lines were ordered dropped to get underway — within minutes, the ship castoff from the dock.

Well-wishers stood alongside waving to friends and relatives as the *Portland's* whistle blew indicating her departure. Overhead, a dismal overcast sky threatened rain and/or the possibility of snow with the temperature hovering at freezing. In lieu of standing on deck fighting the cold wind, passengers preferred the warmth inside the cabin where they could huddle about coal-burning stoves.

Reportedly, 20 minutes later, as the *Portland* passed the Deer Island Light, snow started blanketing her decks. The remaining passengers leaning on deck rails finally took cover inside to hide from the nasty weather.

As later stated by Captain William Thomas of the fishing schooner, *Maud*, it was 9:30 that evening when he first saw the lights of the *Portland* making way on his starboard side. He approximated their position as south of Cape Ann. The captain commented, the weather deteriorated rapidly in high winds and blowing snow. Concerned for his heavily laden schooner, Thomas feared the already ominous looking seas would soon worsen.

Calling to his first mate, he asked to revise their heading. Because of the weather, he decided to take refuge in port at Gloucester rather than continue on into his home port, Boston.

The crew of the *Maud* was the last to see the silhouette of the steamer fade as the storm intensified.

It's estimated, that somewhere between Cape Elizabeth and Cape Ann, gale force winds took there toll on the *Portland*. Apparently, the all-wood steamer was no match for the sharp teeth of the storm.

The ship's arrival was expected momentarily, while family members and friends assembled dockside in Portland, Maine. With the vessel now an hour overdue, a company manager exited the warmth of his office holding a megaphone.

He announced, "Your attention, please! We were just informed, there will be another two-hour delay in docking of the *Portland* due to the bad weather."

As expected, groans emanated from within the crowd. However, most were understanding — the snow continued to fall. This was a joyous time, as many of the ship's passengers were returning from Boston to enjoy the Thanksgiving holiday. Children ran about playfully waiting to see the big ship take its berth dockside.

Several more hours passed. The crowd, now cold and increasingly hostile demanded answers! Someone in the rear of the crowd shouted, "Hey mister, what's going on? Is there trouble aboard the ship?"

Again, the ship's agent exited his warm office.

"Your attention again, please! Folks, please be patient. We believe the ship is delayed due to the inclement weather. That's all we know! As soon as we get more information we'll pass it on to you."

Onlookers groaned again, eventually taking seats in waiting rooms. Children slept on benches, the elderly snoozed quietly where they sat. There was little more to do! Greeters who remained outside fought the nasty, bone-chilling wind and snow as the wait for a hopeful glimpse of the ship coming into port.

Hostility grew, patience weaned. Finally, late the following morning, the office of the Portland Steamship Company reported that their ship, the *Portland,* was now feared lost at sea!

Horrified with the news, spectators stood by waiting for more information. People who had waited all night called out, "You bastards! Why weren't we told sooner?" The company too was in the dark! There were no formal reports or answers.

In a desperate plea for help, calls were placed to officials of neighboring cities asking for possible sighting or the whereabouts of the *Portland*! It was possible that the big ship ran aground in the foul weather. She might possibly have docked at another port to wait out the storm. All aboard could still be safe! Without the availability of wireless communication, no one knew exactly what had transpired that dreadful night!

It was not until the following day that large portions of the ship's debris began washing ashore along the beach at Cape Cod. Sadly, confirmation came. The Life Saving Station stated, the ship's parts were positively identified as those coming off the *Portland.* The vessel was definitely lost at sea!

But why were portions of the ship found in Cape Cod when she originally was sighted by the *Maud* off Cape Ann? Investigators tried placing the pieces of the puzzle together utilizing the facts on hand.

First, Nantucket weather station reported gale storm warnings that night with wind velocity approaching 90 miles per hour. Seas increased, visibility diminished in blowing snow.

Second, they deduced, Captain Blanchard was busy fighting heavy seas driven by strong northerly winds. Her slow forward speed of only 10, possibly 11 knots was no match for the gale-force winds. In reality, she was actually losing ground — steadily being pushed southward by wind and current. It's assumed Captain Blanchard, a well-seasoned master, ordered his engine room crew to produce full steam in order to fight keeping the ship's bow into the wind. By early morning, possibly 3 or 4 a.m., and now eight or more hours into the voyage, the *Portland* may have exhausted her supply of coal, therefore losing power and steerage. Without her ability to steer, the ship broached unwillingly into churning seas.

Next, the speed with which all of this took place might also have accounted for none of her large lifeboats was deployed to save lives. This also could account for no one aboard surviving the sinking!

In all probability, the *Portland* broke up seven to ten miles north of Cape Cod. With its flotsam disbursed over several miles, wreckage eventually found its way to the beach at the cape. Over a period of two more weeks, bodies continued washing ashore. Each was grimly identified as a passenger from the ill-fated *Portland*.

New Englanders demanded a board of inquiry be formed to investigate the tragedy. Adding to the problem, dozens of lawsuits were filed against the ship's owners. Staff members too could only speculate along with investigators as to exactly what had happened. The only eyewitness accounting was the limited testimony brought forth by Captain William Thomas of the fishing schooner.

At first he stated, "In Captain Blanchard's defense, none of us at sea that night had any form of advance notice as to the strength of the storm's intensity. Surely, no one realized it would be that bad! In all due respect, if Captain Blanchard did know, he no doubt would have aborted the voyage north. What more can I say? None of my crew nor I saw the ship go down."

The Ill-Fated *Portland*

Arguments were raised by fellow seafarers brought in to offer their professional opinions. Most of the witnesses offering testimony had fought the devastating storm that night before finally making their way to the safety of another port.

The question was asked, "Why didn't Captain Blanchard too seek another port in lieu of making his destination?"

Still in the witness chair, Caption Thomas merely shrugged his shoulders, he was unable to comment for Blanchard.

With no legal conclusion possible, the court threw up its hands. There were no answers. The closing ruling remarked, "The loss of the steamship *Portland* was contributed as an Act of God."

Spectators left the court confused. Still remaining, was an undertone that the owners of the *Portland* might be covering up the disaster to avoid further legal action!

The New England coast was never a stranger to violent storms, either generating from nor'easters or hurricanes blowing from the southeast. That tragic night, the gale of 1898 was no exception. The storm's rage claimed a total of 71 ships of assorted sizes and descriptions between Connecticut and Maine. The most unforgettable being the *Portland*. It was she who felt the worst of its venomous bite taking all lives aboard.

Sea Search

Lady Liza,
The Little Tramp

The exact date is unclear. Some estimate it was during the late 1920s when the *Lady Liza* left the Norwegian Fjords to sail for the warmth of the South Seas. Captain Helmut Von Zimmer was the ship's master. A rugged businessman who acquired funds inherited as the heir to a Norwegian clothing manufacturer, Helmut disliked the day-to-day drudgery associated with operating a clothing mill. Instead his heart leaned toward the sea.

With a solid cash resource available, Von Zimmer turned his attention from the freezing winters of the north to a life in the open free spirit of the South Seas. He took ownership of an old tramp freighter, hired a crew, then set sail.

With her new owner at the helm, the path of *Lady Liza* moved south through the English Channel, the Straights of Gibraltar, the Suez Canal and finally turning east into the Indian Ocean. Her port-of-call was eventually Hong Kong.

Long before the age of computers and push-button electronics, nefarious explorers discovered the islands of the South Seas. With land distances varying from several hundred to thousands of miles between islands, any form of land was a welcome sight for mariners after sailing for months without land reference. Emaciated seafarers found the tiny islands a supply stop for provisions and water. Seamen soon realized native craftsman were capable of producing high-quality silks, pearls and precious gold jewelry.

Native islanders, considered savages at that time, were taught to trade and work within a white man's society. The less fortunate later found themselves shanghaied, drawn into slavery and shipped back to Europe in chains.

Small trading ships, mostly square or schooner-rigged design vessels, began servicing these islands on a frequent basis. In their holds: firearms, medical supplies and tools were all welcomed by the natives.

Now, after entering the 20th century, Helmut Von Zimmer realized the continuing potential of this trade and he too decided to capitalize on such business.

Operating independently from large mainline carriers, the *Lady Liza* found a virgin marketplace, making port in the Gilberts, the Solomon Island group and, cargo permitting, ventured as far east as French Polynesia. Islanders were thrilled to see the *Lady Liza* drop anchor in a nearby bay while en route to her next port-of-call. Assuredly, Helmut Von Zimmer had chocolate, soda pop and beer in his hold.

The handsome Von Zimmer, a man in his mid-fifties, was often considered a renegade trader, buying and selling goods, then transporting them to whomever he thought would bring the best price. His amoral character was often questioned — not caring who the buyers were, what the item was or where the money came from!

During the late twenties and early thirties, life on most islands was seen as quiet and laid back and of no political value. Von Zimmer enjoyed the leisure atmosphere, warm sea breezes, lucrative trading and pretty young women who fancied his suave European mannerisms. He and his crew would spend days at a tiny port, then move on to the next for new trade. Their return would not be seen again for months.

The ship, later renamed the *Lady Liza* (the original name is unknown) was considered a tramp-freighter, built in Danzig, Germany, in 1900. Originally powered by 160-horsepower coal/wood-burning steam engine, she was later converted to a more convenient Volund Diesel consuming 27 liters of oil per hour. Ship's fuel tanks were large enough to voyage 2,000 miles. Her steel constructed hull measured 110 feet with a 22-foot beam and displaced 150 tons. The *Lady Liza* had a normal speed of 11 knots, carrying a crew of six. Her freight ability provided one rear derrick to load and unload cargo in a single hold. At some point in her career, several portions of the deck house were rebuilt to accommodate paying passengers with comfortable staterooms and lavatories.

While in command, Helmut Von Zimmer was firm taskmaster yet he realized with the limited size of his crew he must pitch in and work alongside the men whenever necessary. Despite his aloofness, the crew held him in high respect.

Lady Liza's crew consisted of the first mate, Otto Schmidt, a former member of the German Navy, who also acted as chief engineer and part-time helmsman. When not at the wheel, he complained constantly about the intensity of the heat generated within the ship's engine room. The tiny enclosure provided only limited ventilation from tropical temperatures. The ship was originally designed to operate in cold waters of the North.

In the galley, was an Italian gourmet cook, who had a questionable background and thought to be hiding from the Italian authorities for murder. No one questioned his past, only his outstanding culinary ability.

Two deck hands were of unknown origin and, later, a young Japanese steward who joined the crew in Hong Kong. The lad served meals, washed dishes and acted as a valet when the *Lady Liza* added paying passengers.

Captain Von Zimmer not only had the responsibility of ship's master, he handled all financial transactions, acted as tradesman, was responsible for the vessel's general welfare and navigation while at sea.

Operating in the Norwegian Fjords, the *Liza* carried a sundry of goods: machinery, livestock, grain, and fuel oil and was normally hired as cheap transportation. If the cargo fit in her hold, she would carry anything, go anywhere!

Von Zimmer continued promoting business the same way in the South Seas as when the *Liza* served northern ports. He was fascinated by her freedom — tramp freighters were noted as having no schedule, no predetermined route and/or destination. Her next port-of-call was based strictly on the load of cargo in her hold and the whim of her master.

The ship could handle up to six paying passengers who shared three so-called staterooms, two flush toilets and one stall shower. Travel was on a first come, first-served basis, all expenses paid in cash prior to departure. Booking passage required only money. Von Zimmer asked no questions, required no passports or security check. Rarely did passenger manifests include the real name of the individual. Most traveled incognito to avoid recognition. The captain often found wealthy passengers who used the ship as a pied-à-terre traveling casually from island to island disguising their identity.

While at sea, life aboard ship was considered pleasant, simple and quiet. Von Zimmer insisted the *Lady Liza* provide good food served three times a day, the best of spirits and clean linen for passenger berths. Little was asked of his guests other they than remain distant from crewmen. Let them do their work!

Female passengers in particular found the ways of their captain charming and loved his rough like mannerism. His handsome Norwegian blue eyes and white beard depicted the image of a Hollywood movie star. Women sat in awe of his debonair appearance. Often, he would secretly accommodate female passengers in his private cabin, asking not to be disturbed for hours on end!

Von Zimmer found virtue in caution when traipsing from island to island. As his ship was only 110 feet in length the small vessel was no match for high seas or the dreaded typhoons frequenting the southern waters. Both captain and crew learned many lessons as they journeyed east into the ominous waves of the Indian Ocean.

One memorable night, he logged:

"Rough seas and high winds are crossing the bow of the *Lady Liza*. On two occasions, as captain, I thought my ship may be swamped by treacherous waves as the water began to fill our cargo hold. We are no match for the violent seas that lay before us."

He attributed survival of the ship to the hard-working crew,

Lady Liza

strong bilge pumps and God's grace.

Although the *Lady Liza* now operated its engine on efficient diesel oil, it proved to be a burden when servicing remote islands. Many had no diesel oil supply to offer which, limited the continuing range of the voyage and the next port-of-call.

It was in the mid-thirties that Japan became a threat to these small atolls. The Japanese needed these islands as stepping- stone refueling and supply stops for their militarized armed forces.

While aboard, the master insisted all hands be familiar with small arms. Although the closing of the decade was considered a modern age, pirates, mostly Oriental renegades, still persisted — hiding in backwater anchorages just as their counterparts did centuries before them. They fought with the same ruthlessness, roaming about seeking small ships that could easily be overtaken and pillaged. Many a ship mysteriously disappeared without trace.

It was late in 1936, while in a restaurant in Bougainville, Solomon Islands, when Captain Von Zimmer had dinner with a fellow captain, Rudolf Bower.

Over a glass of wine, Von Zimmer told his friend the *Lady Liza* just slid down the ways of a nearby dry dock. The ship had completed minor repairs — her bottom scrapped of barnacles and repainted — she was again ready for sea.

He spoke of a morning departure to the East, making their final stop at Bora Bora. Hopefully, his crew would be sober enough to make the early morning departure. Of prime concern was his engineer, Otto Schmidt, who usually spent his days off in the local whorehouse and remained drunk for the duration.

Von Zimmer alluded, his new list of passengers, all of Japanese heritage, were carrying important cargo requiring maximum security and the utmost of discretion. Von Zimmer would not disclose or embellish on the nature of the items only that he was well paid to make the voyage. They had asked their identities remain anonymous to anyone who made inquiries!

Bower reminded Von Zimmer of the hazards while making such a voyage. First, they were in the midst of the typhoon season and that he take special care of the weather and second, be especially cautious of pirates who frequented that area. Japan was now showing aggression in the Orient — invading Manchuria and coastal China provinces. Buccaneer Chinese looked for unsuspecting prey, especially those who transported Japanese citizens. A majority of these renegades were well armed and rarely took prisoners. Those who did survive capture spoke poorly of their incarceration and normally were left to die on a remote island!

Von Zimmer assured his friend this was just a routine voyage and the *Lady Liza* would return safely in a few months. He was well aware of the risks and again commented that his ship was generously compensated for making the trip.

Under cover of darkness the following morning, a steady drizzle fell upon the wharf where the *Lady Liza* was berthed. A heavy fog persisted, blanketing most of the harbor limiting visibility to less than 100 yards.

Two crewmen were seen carrying the partially sober Otto Schmidt aboard the *Liza*. The captain, showing disappointment, ordered him below, then asked the men to take charge of operating the derrick winch. Von Zimmer stood on the fo'c'sle supervising the loading of four large wood crates into the hold.

A black limousine stopped on the pier permitting four passengers to exit, one women and three men. The well-dressed lady seemed to cling to one of the gentleman as though he were her husband or a close relative. After waving off the driver, one man removed a large envelope and handed it to Von Zimmer. The captain peered inside, nodded and welcomed his passengers aboard. Presumably, the envelope contained payment for the voyage. He ordered the young steward to bring their luggage aboard and stow the bags in their staterooms.

A longshoreman, he later testified, he overheard the captain saying to one of his passengers, "Departure would be delayed due to the heavy fog!" The man argued, insisting they leave immediately.

Shaking his head in protest, Von Zimmer mumbled something then took his position in the wheelhouse. Only moments passed when the command was given to the longshoremen to drop the dock lines at the same time signaling the engine room to move astern. The *Lady Liza* backed slowly away from the wharf. Her propeller changed direction then disappeared into the morning fog. That was the last time anyone saw the *Liza* or any member of her crew.

There was much speculation as to how or why the old tramp vanished! Maritime reports stated that three days after her departure, a large unrelenting storm developed to the east. Ships moving through the same area reported high winds, heavy rain and high seas. It is possible the storm may have swallowed the small ship in its wake.

Others made mention that it was common knowledge by fellow seamen who frequented local bars, the *Lady Liza* carried Japanese political defectors who knew of the potential war between their country and industrial nations of United States and Australia. In the wood crates, they no doubt carried valuable paintings, art treasures and gold to be hidden on Bora Bora or a nearby port. The vessel was a treasure ship overtaken by pirates, looted, then sunk undetected!

It was learned, Otto Schmidt, the ship's engineer, only hours before the sailing, became drunk at the bar in a local brothel. In a boisterous voice, Schmidt bragged to several prostitutes accompanying seamen he was about to be a rich man! The next voyage would be his last! His ship was about to embark on a secret trip to Bora Bora. When he returned, he was going to retire to his home in Germany.

According to Rudolf Bower, the man who had dinner with Von Zimmer the night prior to departure, "In my opinion these people who came aboard must have been Japanese officials fleeing the country. I believe these individuals were running due to possible exile by the Japanese or even Chinese government! Surely they were people of importance: if not, why was there so much secrecy to the voyage? Why was there such a high price paid?"

He added, "Many knew of the *Lady Liza's* departure to the east.

I'm positive radical sympathizers to Japan's cause were the ones responsible for the loss."

Whatever the reason for her disappearance, the little tramp freighter, *Lady Liza*, will long be remembered by the South Sea islanders as the welcomed supplier of their needs.

The Ship That Returned From The Dead

In the vast panoply of World War II the loss of a single naval trawler was just another grim statistic to be added to a long list of ships lost at sea. The casualty rolls were filled with thousands of ships of every description, type and tonnage lost in the worldwide holocaust and the addition of a lone naval trawler of 310-tons gross with a 21 man crew caused little more than mild notice and some additional clerical work on the part of naval personnel who had become inured to handling such tragic matters.

So it was in mid-January, 1944, when the bantam-sized 123-foot trawler *Strathella* simply vanished from the sea following a hurricane-force gale several hundred miles off the northwest shores of Scotland. En route to Iceland as one of three Royal Navy Patrol Service trawlers escorting a small convoy of four merchant ships, the *Strathella* was last seen by the escort commander, Lieutenant R.V. Downer, RNR, as she surged through 30 foot high combers struggling to keep position on the convoy's rear flank. The gale continued to pound Convoy UR-105 for 48 hours and when it finally blew itself out there was no sign of the *Strathella* anywhere. Steering his 650-ton trawler, *Northern Spray*, close to the second escort, Downer anxiously inquired over the loud hailer if anything had been seen of the missing *Strathella*.

As he feared, the other skipper's somber reply was that he caught only a brief glimpse of her through the flying spindrift and driving rain in the early stages of the storm. Further inquiry to the captains of the plodding merchantmen produced no evidence or clues to the trawler's whereabouts, for at the height of the storm, visibility had been down to less than a quarter mile.

Concerned about the fate of his missing convoy-mate, Downer decided to risk breaking radio silence by calling *Strathella* on the wireless. Repeated voice calls brought nothing but the crackle of static on the loudspeaker. Nor were any reassuring dit-dahs of Morse code heard in the radio room of Downer's *Northern Spray*.

Not daring to leave the convoy unguarded, all the two remaining trawlers could do was sweep the horizon for signs of the missing ship. By nightfall Downer felt a nagging fear deep in the pit of his stomach. The 38-year-old *Strathella* had been plagued by a balky engine ever since they set sail from Loch Ewe. Her 30 year-old skipper, Temporary Lieutenant Osmund Lee, RNVR, while showing great aptitude for the sea despite the fact he had been a London accountant only three years earlier, might not have had the experience to keep his vessel from foundering if he lost his engine at the height of the storm. All Downer could hope was that somehow the *Strathella* had survived the storm and would soon turn up in Iceland, their destination.

But each passing day made the 1906 vintage trawler's chances of survival that much more remote. It was known her coal bunker capacity was limited to 60 tons and that at normal eight knots cruising she consumed over eight tons of coal a day, or had a cruising range of about a week.

When the convoy reached Reykjavik on January 19, a week after the storm had hit, it seemed certain the *Strathella* had gone down, for the shore base had heard nothing from her. Downer promptly reported her disappearance to the Admiral Commanding and the admiral immediately ordered an extensive air and sea search though the wintery weather was still very bad.

For four days, PBY patrol planes and sea-search craft combed the route the convoy had taken. Late on the afternoon of the 23rd, a PBY crewman spotted a half-sunken lifeboat awash on the sea. Banking steeply for a closer look, the crew saw the weather-worn name *Strathella* on the boat's bow. There was no sign of life

Returned From The Dead

anywhere. A day later the search was called off and Admiral Commanding Iceland informed the admiralty the H.M. trawler *Strathella* was still missing and must now be presumed lost. She was at that point well over five days beyond the range of her fuel.

Doubtlessly, the aged trawler, armed with only two six-pounder guns, depth charges and two Browning machine guns for air defense, had gone down in the terrible gale of January 13-14.

Soon after, word reached the admiralty: "*Strathella* — Lost With All Hands" a routine chain of oft-repeated events rippled through the casualty section that would soon see the families of 22 officers and crewmen notified that their loved ones were missing in action. With far less emotion, clerks in the administrative and accounting office began the dreary paperwork that would remove the luckless trawler from the active ships list and consign the mounds of records of her being to the overloaded shelves of the war records section. The *Strathella* simply ceased to exist and all memory of her, save for the broken-hearted families of her lost crew, vanished and was forgotten in the fast-moving pace of the war at sea.

But unbeknownst to anyone, the tiny trawler was still very much alive. As her valiant crew fought the Arctic cold and vicious storms, they refused to give up their fight for survival even though their crippled ship had been aimlessly drifting ever closer to the frigid death-trap of the Arctic Circle.

For the best part of the 48 hours that the gale of January 12-13 continued to rage the *Strathella* had been compelled to heave to. Rolling alarmingly, with heavy 30-foot-high seas breaking over her, the shuddering ship somehow managed to take the wind-whipped blows assailing her tough hide. Though skipper Osmund Lee was a neophyte mariner, a "hostilities only" volunteer, he'd learned his new trade well in his three years in the Royal Navy and remained on the wave-swept bridge beside the helmsman throughout the storm. Fighting his own fears, he calmly gave each required order as he sought to keep the ship headed into the mountainous seas. Already one giant wave had wrenched both bridge machine guns from their steel stanchions. As they swept overboard they had shattered the 10-inch signaling searchlight on the port side.

Not long after, another wall of water tore their solitary 16-foot skiff from its davits, carrying it away in a frothing trail of flying spindrift like so much useless flotsam. With each pounding wave it was anyone's guess how much more the battered ship could take. She'd been built well in Aberdeen in 1915 and her steel hull had weathered many a storm in her long service with the fishing fleet. Taken over by the Royal Navy in 1940, she had been employed as a Harbor Defense Patrol Craft at Akureyri, Iceland. Armed with two six-pounder guns, one on the forecastle and the other on a sponsoon immediately before the heightened bridge, these plus the Browning machine guns and added weight of the depth charge racks on her fantail made the tiny vessel extremely top heavy: a cause for alarm even in moderate seas.

Just after midnight on the 13th, Lee's executive officer, number one, in the Royal Navy, struggled to the bridge and breathlessly announced that two depth charges had broken loose of their lashings and were wildly careening along the aft deck. Ordinarily, Temporary Sub-Lieutenant Alan Bateman was an officer who remained nonplussed in any situation. But as he shook the water from his storm-drenched slicker, Lee could see the fear in Bateman's eyes.

Mustering his own inner strength, Lee calmly asked Bateman to call for volunteers to attempt to secure the explosive charges. It would be a hazardous task with the deck pitching and rolling beneath them, but Bateman drew in his breath and said he would see what he could do. A long half-hour later Bateman returned to the bridge and gave Lee a confident thumbs-up. The depth charges had been secured and none of the four volunteer matelots had so much as crushed a fingernail.

But the fury of the storm was unrelenting and as the first eerie light of dawn ebbed through the Stygian darkness another mountainous wave staggered the trawler and burst open the door of the radio shack. The cascade of frothing water lashed at the radio transmitter, short-circuiting the wiring and blowing out the transformer in a shower of sparks.

The storm-swept ship was now deprived of its only source of communication with the other vessels. As it was, Lee had no idea of the location of the convoy or their proximity to it. His first and only concern was keeping the *Strathella* afloat and that in itself consumed his energies.

By mid-morning the gale began to abate and Lee anxiously attempted to plot their position. But the radio direction finder was smashed beyond use and a heavy overcast prohibited taking a sun sighting. With a heavy rain still splattering her decks, a whole new series of woes overtook the bantam warship.

Engineman Rudy Harkness, and his colleague James Heap, both veteran trawlermen who could be counted on in any situation, wearily announced that the tired and overstrained engine had finally broken down. More than a score of the boiler tubes were badly leaking and steam pressure had fallen dangerous low. Making matters worse, the engineman grunted dejectedly, if the engine could be fixed they had to face the fact they were now almost out of coal, for in fighting the storm they had consumed more than the normal amount of fuel. Harkness reckoned they had enough left in the bunkers for a bare three days cruising at best.

Skipper Lee listened attentively, then outlined his plan. The first priority was to fix the engine. Second was to begin rationing the food and water they had aboard and, third, they'd rig a sail that might help them find a landfall. With Bateman in the wheelhouse, Lee then candidly announced he was going to his bunk. He hadn't slept in almost four days.

Miraculously, engineers Heap and Harkness got the main engine running once again. But it taxed every ounce of their ingenuity, for in the height of the storm many parts had actually fallen off and had to be replaced with seizing wire, tape, rope and whatever bits and pieces could be salvaged from the storm's mayhem.

Wireless operator Charles Stropshire and ASDIC sound man

Sea Search

Hugh Nellis put their heads together and suggested that they try to repair the burned-out wireless by scavenging parts from the ASDIC sound detector, an early type of Sonar used to locate submerged submarines. With Bateman's help they worked for three days cannibalizing the units, but after all of their efforts the signal Stropshire transmitted was so weak it could only be heard a few miles away. Then the last critical tube (a British valve) burned-out, permanently cutting them off from all communication with the outside world.

The spanker sail fashioned from the canvas boat covers did little to keep the *Strathella*'s head into the wind for the high bow blanketed the air leaving the sail to flutter uselessly in the breeze. Skipper Lee suggested a larger trysail be made from every possible material that could be found. The sail would be crucial, he informed the crew, for by then, January 21, barely three tons of coal remained in the bunkers.

Because of the bitter cold Lee wanted to save what coal was left for heating and cooking as each day saw them aimlessly drifting closer to the barren wastes of the Arctic Circle.

The older crewmen knew Lee's deepest fear that the powerless ship would haplessly drift into the crushing ice floes north of them. As experienced fishermen, they knew of the Irminger current that flowed cyclonically from the landmass of Iceland. It moves westward where it joins the freezing floe from the Arctic. Thus, the Irminger current had first carried the trawler away from Iceland and into the orbit of the East Greenland current, which was now taking the vessel northwestward at a speed of some five to ten miles a day.

The days slowly dragged by with nothing to be seen but the empty expanse of the North Atlantic stretching around them. By then Lieutenant Lee knew that each passing day made their chances of survival that much more remote. Any search activity would have been called off for the *Strathella* was now well over two weeks overdue.

In the minds of even the most optimistic comrades ashore the luckless trawler had either succumbed to the storm or had fallen victim to a U-boat. Her limited fuel capacity was well known and being essentially a harbor craft pressed into convoy duty, *Strathella* carried only the barest of rations, fresh water and spare parts. Indeed, with no landfall in sight, Lee ordered rations further cut from two meals and two cups of tea per day per man to one meal per day and a half a pint of water per man.

Maintaining a cheery confidence despite his own inner doubts, Lee ordered the shipboard routine to be as rigidly maintained as if they were on a normal patrol. Keeping the crew busy was their only way of not panicking over their plight. Somehow, someway they'd reach a landfall or be found, Lee insisted. Buoyed by his self-confidence, the crew went about their chores of cleaning the guns, maintaining watches, making reports and keeping a watchful lookout.

Early on the morning of January 27 an excited yell from the lookout brought everyone topside. A spit of land was spotted 30 miles to the northwest. Without proper charts Lee could only guess that they were somewhere off the east coast of Greenland. Ringing the engine room, he called for all the revs the balky engine could deliver. Bringing up the power meant squandering what precious coal was left, but Lee reasoned it was a risk he had to take. Turning to his number one, Bateman, he ordered every possible scrap of wood to be torn from the decks and fed to the ravenous boilers.

Late that afternoon their hopes of reaching land were dashed when they found ever-thickening fields of ice blocking their path. Lee kept probing and weaving hoping to somehow find a clear channel, but each turn only proved to be a more impassable barrier than the last. With only a few hours of steam left in the boiler, Lee reluctantly ordered the ship turned about and headed for the safety of the open sea.

Shortly after sunset the smokestack belched its last puffs of inky fumes and the engine ceased to turn. As a freezing cold set in the *Strathella* again resumed her heart-breaking drift westward into the numbing void of the Arctic.

As the days passed drearily by, Lee kept the crew's spirits up by reminding them that since they'd spotted land once they would find it again. But by then they were only a few degrees from the Arctic Circle and each day saw a thickening mantle of ice entombing the ship's decks, superstructure, gun platforms and rigging. But even this hazard proved to be a blessing in disguise, for chipping away at the icy tentacles kept the men busy and blood circulating in their numbed flesh. The ice also provided them with fresh water, a vital commodity that was almost exhausted.

With their last food becoming a dimming memory, Lee silently admitted their plight was desperate. If something wasn't done to stop their endless drifting the ship would soon become a ghost vessel manned by the skeletons of 22 hapless seamen. Once again he called for the making of a sail that would give them some steerage. The half-frozen crew mustered their last energies and, using the remnants of the original spanker, plus bedcovers, mattress fabric and blackout shades they fashioned a sail large enough to catch the wind effectively.

The tedious work of sewing and scrounging took four days to complete during which time they were besieged by more bad weather. Icy squalls pelted the rust-streaked hull and superstructure as wind-whipped snow blotted out all trace of the horizon. Half-starved and half-frozen, everyone aboard helped finish the patchwork sail: their last hope of making a landfall or avoiding a collision with the ever-more-present icebergs.

At dawn on the 12th of February, the sail was finally hoisted. The weak and weary crew cheered as a gentle breeze filled its taut expanse. Soon they could feel the ice-laden hull begin to move as helmsman Alan Anderson, joyously took up a course for the first time in weeks.

Hours later the lookout mustered his strength to cry out that he'd spotted an aircraft high overhead. Lee had drilled his crew time and time again on what to do in such an event and within seconds a flurry of flares and rockets shot into the gray sky.

On the bridge, the signalman flashed away with the Aldis lamp. Then their spirits crashed as the plane droned away, never altering its course.

Ever resourceful, Lieutenant Lee quickly made the most of the spotting by reminding the crew that the aircraft must have a base somewhere nearby for it flew at an altitude lower than normal for that region. He was convinced they'd drifted into a regularly patrolled sector and immediately ordered the lookouts doubled.

The next day Lee's optimistic assumptions were given some credence when dawn revealed the *Strathella* to be only four miles from a definite shoreline. Edging closer, they soon discovered the ever-present icebergs blocking any course as they drew shoreward. Knowing the effect of his next order, Lee once again ordered the tattered trysail hoisted so they could steer clear of the ice and not risk puncturing the hull.

Faces grew grim as once again the bow swung away from sight of land and headed seaward. Were their strength not so weakened Lee feared they were at the point of rebellion. But somehow they resisted the desperate urge to abandon ship and try to make their way across the treacherous pack ice. Almost at his own wit's end, Lee slowly drew in his breath and asked Bateman to take over the bridge. He couldn't bear to face the next decision he'd have to make.

Alone in his cabin he quietly contemplated their fate. No matter which course he followed they seemed doomed. The open sea would keep the ship intact, but in a few days hunger and cold would begin taking its deadly toll. To try and leave the ship and cross the ice was rife with hazard for without a boat they had no way of crossing any open water that might be encountered in the ice fields. And, if they did reach land all they might find was barren ice and snow-covered terrain that might be impassable to cross one foot. No, staying with the ship was still their only hope, he reasoned.

Late in the day, as a blue-cast sun sank beneath the horizon, the lookout again screamed that a plane was in sight. Forcing their bodies to shrug off the numbing cold, the crew scampered topside and once again fired salvoes of rockets and flares skyward. For a second it looked as if this aircraft also would continue unseeing on its course. Then, suddenly, it banked to the left and headed toward them. Waving and cheering, the crew became joyous for as the giant seaplane zoomed over their ice-crusted mast they saw the star and bar insignia of a U.S. Navy PBY patrol plane.

The signalman's lamp winked away in a frenzied flurry of flashes that announced their name, port and condition. Sweeping around them in a wide circle the PBY flashed back that their message was understood and help was on the way.

Soon after dawn on the following day, February 15th, the USCGC *Modoc* hove into view and soon pulled alongside the woebegone *Strathella*. As skipper Osmund Lee greeted and waved at the cutter's captain from the bridge, his whiskered and drawn crew stood at mute attention paying silent homage to the landlubber accountant from London who had seen them through their long ordeal.

His Majesty's trawler *Strathella* had returned from the dead thanks to Lieutenant Osmund Lee's indomitable courage and leadership.

The Treasure Ship
Brother Jonathan

Captain De Wolfe was so furious he had to clench his fists to control his temper. A man of strong principles and rigid ethics, he could not fathom how the agent could condone such outrageous overloading of a vessel when the lives of so many hapless passengers were at stake. The argument between the tall, lean, bearded captain and the bellicose agent raged on, each shouting angry condemnations at the other above the whistles and horns of the rain-swept San Francisco wharf.

The captain's point was that the steamer *Brother Jonathan* could not safely run the storm under such grossly overloaded conditions. As it was, the *Jonathan* already boasted an unsavory past, having been often plagued with breakdowns of its cranky engine and poorly maintained boilers. Built for the calmer waters of Long Island Sound in 1851, the creaking side-wheeler had nearly foundered more than once in her 14-year career; the last incident occurring only a few years earlier off the Washington coast with 370 passengers aboard. But the agent's argument was equally hard driven; any captain worth his salt wouldn't be afraid of a slightly overloaded vessel. The owners — in this case the California Steam Navigation Company — wanted profits and the only way to achieve them in the cut-throat shipping business was to cram every inch of space in every vessel with paying cargo — human or otherwise.

Captain De Wolfe's entreaties met with abject refusal. Scowling, the heavy-jowled agent shook his fist and bellowed for all to hear if De Wolfe didn't obey orders, he would find a captain who would. The argument was settled, for the agent was also the brother-in-law of one of the CSNC company vice-presidents. It was a "take it or leave it" situation. Captain De Wolfe, perhaps out of a sense of pride, decided to take the run just as he always had in the long years past.

Pulling up the heavy collar of his peacoat, he squinted into the driving, wind-swept July rain and watched the cargo hoists groan and strain as the loading continued. Shaking his head, he wished in retrospect that he had decked the viciously arrogant little agent. It would have cost him his job, he considered, but he knew he enjoyed a good reputation as a capable, astute seafarer who rode the narrow line between caution and daring at sea as effectively as any skipper on the waterfront.

What De Wolfe didn't know was that his decision to sail into the teeth of a raging summer gale would cost the lives of more than 200 people.

The tragedy would become the worst disaster in California marine history and the whereabouts of the sunken *Brother Jonathan* would haunt treasure hunters for generations to come. Though the exact amount of gold on board can only be speculated upon, it is known the vessel was carrying a payroll of more than $300,000 destined for garrison troops in the Pacific Northwest. In addition, Wells Fargo gold and silver bullion in transit to Victoria, British Columbia, the *Brother Jonathan's* destination, was rumored to be in excess of one million dollars. But the payroll and the bullion would never reach their destinations, for fate would intervene somewhere offshore along the treacherous St. George Reef.

The following day, San Francisco harbor bustled with activity even as the summer 1865 storm continued to whip the broad waters of the bay into frothing white-caps. At the dock, a steady stream of passengers poured aboard the 1,360-ton vessel. Among them, much to Captain De Wolfe's disdain, were a dozen highly rouged scarlet ladies — prostitutes from San Francisco's bawdy Barbary Coast who were moving north to ply their nefarious trade in the booming Pacific Northwest where lumberjacking, mining and prospecting

Brother Jonathan

made lonesome men eager for a female's soft caress. The Civil War was over, the states were reunited once again — and coastal shipping was thriving to the point that ships had to leave cargo on the docks. The Gold Rush had run its course, but brisk coastal commerce from the timber rich redwood country of the Humboldt to the Silver Gate at San Diego kept anything that could float busily occupied.

Watching the last of the passengers board the creaking gangway, Captain Samuel B. De Wolfe felt an uneasy twinge deep in his intestines. A religious man, he disliked having harlots aboard his vessel. The rest of the passengers were the normal mix of people moving north and south, some on business, some to resettle, some on holiday. Though many were ex-Johnny Yanks and Johnny Rebs eager to start new lives, they were, by and large, sincere working men who didn't need the temptations of the flesh so blatantly flaunted by the Barbary Ladies of the Evening. Scarlet women were a bad omen, the captain thought quietly, and he for one would enjoy seeing how these boisterous, over-painted women, chaperoned by Madam Sophie Keenan, would take to the heavy seas swirling beyond the Golden Gate.

As for the rest of the passengers, many of them families or women and children sailing alone up the coast to stops at Portland, Oregon, and Fort Vancouver, he hoped most would have the good sense to remain in their cabins and not risk the hazards of angry green water cascading along the decks.

De Wolfe also noted several well-known celebrities on the manifest, among them General George H. Wright, a grizzly, feisty campaigner from the Mexican and Civil War. The general was traveling with his wife and a small contingent of soldiers guarding the payroll destined for Fort Vancouver. In addition, the general was taking two camels up north to see how these desert animals would fair in a colder climate. The camels shared stalls with the horses of his mounted escort. Another notable was newsman James Nesbit, of the *San Francisco Evening Bulletin*, a writer distinguished for his descriptive prose. With the ship preparing to sail, Nesbit gathered in the crowded salon with several prominent Bay Area citizens.

As the captain watched the last of some 600 barrels of whiskey being loaded into the forward hold he took mental note of the passenger manifest. The list showed 168, but another two dozen or so were still buying tickets from the purser, as was common practice in those days on coastal runs when last minute arrivals were routine. His four Francis Patient Metallic Lifeboats, two smaller surfboats and 400 lifejackets were more than adequate in the case of emergency, he considered thoughtfully.

At the stroke of noon on July 28, the *Brother Jonathan* cast off her lines and slipped out into the stormy Pacific, heading north. A persistent headwind and a heavy sea, plus the overburdened condition of the steamer, made progress difficult. Many passengers were confined to their staterooms and much of the gaiety usually attendant on the early coast-wise passenger steamers was noticeably absent. It wasn't until July 30 that the vessel passed Crescent City, barely holding her own against adverse headwinds.

Captain De Wolfe doggedly kept his course until 1:00 p.m. when the vessel gained a point 16 miles northwest of Crescent City. The gale had reached full proportions. Here, all forward progress was thwarted and the skipper's only alternative was to come about and seek shelter until the storm blew itself out. The vessel had logged about six miles in the opposite direction when suddenly she struck something with a fearful jolt. The passengers stood frozen in their tracks. The steamer was hung up on inundated Northwest Seal Rock, part of the dreaded St. George Reef, eight miles from the mainland.

Captain De Wolfe quickly ordered the engines reversed. Groaning under the strain, the paddles splashed and thrashed into the pounding waves, but the *Brother Jonathan* refused to budge. Cursing his luck, the callous agent who had overloaded his ship and the scarlet women who bedeviled him, he demanded still more steam. Once again the engines bellowed their might. Suddenly the ship seemed to break free of the vicelike grip that held her fast. The shaken passengers had hardly breathed a sigh of relief when another surging wave smashed against the wallowing hull, careening the ship even harder against the unseen, rocky pinnacles. This time the blow was fatal, for all too soon pieces of broken keel came floating to the surface. The ship was impaled, her back broken on the treacherous, jagged rocks beneath the surface. The second bone-chilling lurch in the gale-swept sea caused pandemonium to break out. Frightened passengers scrambled over each other for the boats as the *Brother Jonathan*'s shooting steam whistle blew its last gasps for help.

The only survivor on watch at the exact moment of the crash was quartermaster Harold J. Yates, an Englishman who had served in the Federal Navy and had been in the Battle of Mobile Bay. He later would give this testimony:

"I took the wheel at 12 o'clock. A northwest gale was blowing and we were four miles above St. George. The sea was running mountain high and the ship was not making headway. The captain thought it best to turn back to Crescent City and wait until the storm had ceased. He ordered the helm hard a port. I obeyed, and it steadied her. I kept due east. This was about 12:45. When we made Seal Rock the captain said, 'Southeast by south.' It was clear where we were, but foggy and smoky inshore. We ran till 1:50 when she struck with great force, knocking the passengers down and starting the deck planks. The captain stopped and backed her, but could not move the vessel. She rolled about five minutes, then gave a tremendous thump and part of the keel came up alongside. By that time, the wind and sea had slewed her around until her head came to the sea, and she worked off a little. Then the foremast went through the bottom until the yard rested on the deck. Captain De Wolfe ordered everyone to look to his own safety, and said that he would do the best he could for all."

With few bulkheads to stem the rush of water flooding her hull, the stricken ship began to list heavily and go down by the bow. Frightened, witless seamen scurried from the flooding engines shouting that the boilers were certain to explode as sea water sizzled into the burning coals. Flickering oil lamps extinguished as the ship pitched and rocked on her agonizing deathbed, plunging the narrow passageways into Stygian darkness. Children screamed for missing parents and women sobbed their pitiful pleas as loved ones were lost in the maddening rush to get to the boats.

From that moment, confusion reigned. The lifesaving apparatus was unfortunately in serious need of repair. Great holes had been opened in the ship's hull. The gale-whipped seas slammed at the port quarter veering the vessel around until it came head to the wind. The wedge-shaped pinnacles holding her captive ripped open the bottom of the ship and the foremast virtually dropped out of sight.

The pushing, screaming crowd gathered about the lifeboats as harassed crewmen tried desperately to get them lowered. One boat was finally cleared away, but so many of the hysterical passengers piled into it that it capsized. All were drowned. This made it difficult to get passengers into the second boat. When it was almost filled, the steamer took a severe lurch and the craft still dangling in the falls plummeted into the water and swamped.

Ten of the 12 Ladies of the Evening from the Barbary Coast perished in that second boat, along with a number of other women and children. Lines were thrown into the seething water for survivors to grab onto, but in the fury of the storm and the icy darkness, the lines were pulled in empty. Though many frantic, screaming voices could be heard above the wailing wind, mercifully few of those on the sloping decks of the steamer could see the torment and wild-eyed horror of their drowning wives and children. The Grim Reaper was having a field day off the rain-swept California coast.

All at once it seemed as if Captain De Wolfe's grim foreboding about the ship's fate was tragically coming true. The rumors and tales about the *Brother Jonathan*'s stability and seaworthiness, which had dogged the vessel much of her working life, were coming true. Yet, there was little that was really negative about the sidewheeler when all was brought to account during the later investigation.

Built in New York by a capable ship yard in 1851, she was sold into Pacific service with Vanderbilt's Nicaragua Line. Vanderbilt later sold the still near-new steamer at a handsome profit to William T. Wright who renamed her *Commodore*. In olden times changing a vessel's name was considered an ill-omen by many seafarers and so it was that the *Commodore* nearly foundered off the Washington. coast after making her new owner a veritable millionaire.

Perhaps Wright was spooked by the ship's reputation or saw the handwriting on the wall, for he wanted no part of a disaster in dire need of repair following her narrowly averted brush with death. The *Commodore* was sold to her last owners, CSNC, who — perhaps in hopes of changing her luck — renamed her the *Brother Jonathan* once again. Her new owners spent considerable time and money refitting the steamer and before long she was back at sea making a small fortune for the happy new owners. Unfortunately, in those wild and woolly days there were few laws to protect innocent passengers who most often trusted to blind luck that they would make their destination safely. This joyous happenstance would not befall many of the 163 hapless souls aboard the fast-sinking *Brother Jonathan*.

The third mate, James Patterson, who had been asleep when the ship struck, was now on deck attempting to get still another lifeboat over the side. He succeeded in placing five women and three children in it, but before he could round up additional passengers, ten members of the crew leaped in. Patterson was not in the stem sheets and against his orders, the crewmen lowered the boat. Whether through a miracle or the capabilities of officer Patterson, this boat carrying 16 occupants was the only craft that reached shore and in it were the sole survivors of the wreck.

None could ever know the grim and horrible moments as the *Brother Jonathan* was swept from the reef, taking down with her nearly 200 souls. All but 66 were passengers, many of them women and children.

The coastal steamer *Del Norte*, skippered by Captain Hank Johnson, a 20 year veteran of California waters, was dispatched to the scene after the overdue *Jonathan* was reported lost by the handful of survivors who had made the rocky shore. Johnson rescued the 16 stranded survivors from the beach near Crescent City and commenced an extensive search of the area. Though several bloated bodies were found at sea or washed up on surrounding beaches, no other survivors were encountered.

The handful of survivors were not anxious to tell of their harrowing experiences. They remembered all too vividly those last terrifying moments, with passengers gathered about Captain De Wolfe pleading for help that was beyond his power to give. The captain's last words were like the voice from the tomb. "Tell them," he said, "that if they had not overloaded us we would have got through all right, and this never would have happened."

Among the stories that pulled at the heartstrings concerned journalist James Nesbitt, who, awaiting his death, sat down on the ship's forward hatch cover to write his will in his little notebook. He then wrapped it neatly and tied it about his waist. After the steamer went down his body was cast upon the beach, the notebook still secured about him.

And there was Brigadier General Wright, who made such an inspiring tableau as he placed his coat about his wife. She was to have gone into the one boat that gained the shore, but her devotion to her husband was a binding tie. They sank beneath the waves, arms wrapped about each other.

It wasn't long before 78 bodies washed ashore as grim reminders of that tragic evening. Many were mutilated beyond recognition and buried in a mass grave. Others, like some of the army veterans who had so recently fought at Shiloh, Manassass and Vicksburg, were buried with full military honors. Though few of the bodies of the scarlet ladies were found there was deep mourning in the fun palaces of the Barbary Coast where they had frolicked and danced so happily only days earlier. The body of Captain Samuel De Wolfe was never recovered. In the inquest that followed weeks later little blame was placed on him for the tragedy. The captain had followed orders and gone to sea against his best judgment in an overburdened vessel. The ship's agent who had overloaded the ship took no responsibility for his actions, charging instead that overloading was commonplace and who was he to say when a ship was too full above or below decks.

In the way of all tragedies afloat or ashore, the *Brother Jonathan* disaster would soon be forgotten in the long wake of never-ending sinkings off the treacherous California shoreline. Within 10 years a worse disaster would grip the headlines when the *Pacific,* a big side-wheeler, sank with a loss of all but two of her 279 souls aboard following a collision off Cape Flattery Light in 1875. In 1918, the S.S. *Sophia* would sink with the heaviest loss of life ever recorded on the Pacific coast. But the *Pacific* and *Sophia* sank in the waters of the Pacific Northwest leaving the hapless *Brother Jonathan* with the dubious distinction of being California's worst maritime disaster.

Years later Civil War hero Brigadier General George Wright, who went down with his wife aboard the *Brother Jonathan,* would be honored by having an army fort named after him in Spokane, Washington.

More than two decades after the disaster that claimed 203 lives occurred, a powerful lighthouse was erected on the rocks above St. George. For generations to come its powerful lantern beam would sweep the seas above the ghostly remains of the *Brother Jonathan.*

Many expeditions searched for the *Brother Jonathan*'s bones. The first quest began on August 20, 1865 with the steam tug *Mary Ann* as she proceeded north from San Francisco. She stopped at Trinidad to investigate some wreckage. When her crew weighed anchor, they discovered a rusty chain attached. Subsequently it proved to have come from the barque *Acadia,* sunk there in 1861. Their search for the *Brother Jonathan* ended in failure.

In 1894, a man boasting to be the world's "Champion Deepsea Diver" with letters of introduction from Captain James Colton of the Diver's Union, New York, announced his purchase for $200 of the ship's location from C.W. Hill, assistant engineer of the wrecked ship. He claimed he easily found the wreck and not only walked across her deck, but picked up several brass buttons from the drowned soldier's uniforms.

The mysterious diver said the wreck lay in only 200 feet of very calm water and in such excellent condition that he planned to raise her with the aid of three huge air bags, then put her back into service! Local citizenry laughed to the boaster and his remarkable scheme out of town. Even in 1894, public gullibility had its limits.

The bell of the *Brother Jonathan* sits on display at the Light House Museum in Crescent City, having floated ashore there attached to other wreckage. A large wooden eagle that once adorned the wheel box also washed ashore. The salvager sold it to a local bar owner for $60. For years it gazed down at drinkers from a prominent position behind the bar. Presently, it resides at the San Francisco Maritime Museum.

And so the intriguing legend of the *Brother Jonathan* lives on, still luring adventurers to discover the whereabouts of her rotting remains. A disaster that occurred only 15 years after California achieved statehood continues to fire the imaginations of salvagers eager to write the last chapter of this tragic vessel's demise.

The Salvage Miracle At Pearl Harbor

Fire, smoke and explosions still thundered and wreathed Pearl Harbor as Navy and civilian salvage crews began the awesome task of getting the devastated fleet back into action. What mattered now was not why the attack had been so successful, but how quickly the many ships could be made battle worthy again. Accomplishing this feat was the task of salvage crews and they met the seemingly impossible challenge with a dedication and resourcefulness that marked their efforts as the most outstanding salvage achievement in modern times.

The first priority was rescuing trapped crewmen still alive in the sunken and twisted wreckage of many ships, especially the capsized *Oklahoma* and *Utah*. Tappings from inside the overturned hulls told rescuers in Morse code that many still survived in flooded compartments. Quickly gathering the necessary cutting torches, air compressors and other salvage equipment, rescue teams hurriedly cut away steel plating to free the suffocating men. In all, more than 100 lives were saved in these frantic efforts, 32 on the *Utah* and *Oklahoma* alone.

The formal start of the Salvage Organization began a week after the attack with Commander James M. Steele, skipper of the *Utah*, as provisional commander of the hastily garnered force. Gathering teams of divers and salvage experts from every one of the 100 ships still in the harbor, Steele set about the staggering task of finding the necessary equipment to commence operations. In late 1941 Pearl Harbor was still a relatively new base, barely three years old, and supply shortages existed everywhere. Scrounging became the order of the day; and with deft cunning crews, began demonstrating their skills at obtaining items like heavy-capacity pumps, vast quantities of timber for shoring and cofferdams, cutting torches, welding rigs, piping, hoses and hundreds of other items necessary to refloat and repair the battered hulls of ships.

Washington placed the highest priority on getting as many ships as possible ready for sea in the shortest period of time. No one knew where or when the Japanese would strike again. With so many major ships badly damaged and 80 percent of the Navy and Army Air Force knocked out, the weakened defense line needed anything afloat it could muster.

Combining the three major commands of Fleet Battle Force into one headquarters at the undamaged submarine base expedited coordination of the repair effort. The first ships tackled were those least damaged and thus easiest to get back in action. Largest of these was the lightly damaged *Pennsylvania* which was in dry dock at the time of the attack. Bomb and fire damage was quickly repaired with the battleship able to leave the dry dock by December 12 and the Navy Yard by December 20, 1941.

Clearing the flooded dry dock in which *Pennsylvania* was berthed was mandatory since it was vitally needed to facilitate repairing the hull damage of the other wrecked ships. Working around the clock with all available naval and civilian manpower, salvage teams soon had the bow-flooded *Honolulu* in the dry dock vacated by the *Pennsylvania*, where she remained until January 2.

As work progressed, contractors from the Pacific Bridge Company rushed to complete construction of the nearly finished second dry dock they were building when the attack occurred. Obtaining suitable large bilge blocks from the harried Navy Yard, they managed to get part of the dry dock in operation three days after the raid to order to begin repairs on the cruiser *Helena,* which had been hit by a torpedo on the starboard side. By the 5th of January, *Helena* was underway with half power for Mare Island, California.

The Salvage Miracle At Pearl Harbor

Although no lesser personality than Fleet Admiral Chester Nimitz, who took command December 31, thought the fleet largely irreparable, ships began returning to duty or back for mainland refits with remarkable speed. Crews on the battleship *Maryland*, hit by two 15-inch armor-piercing bombs, effected repairs without dry docking once high-capacity pumps were found and was back in action by December 20, two weeks after being considered mortally wounded. The same happened to the aging *Tennessee,* which had been moored inside of the badly hit *West Virginia.* Considerable effort was needed to free the huge warship trapped between the quay and the sunken *WeeVee*, but by December 20 *Tennessee* floated free, her bomb-blasted forward turrets repaired, and the ship ready for duty.

Similar all-out efforts got the cruiser *Raleigh*, hit by a torpedo, the repair ship *Vestal*, beached to prevent sinking, the seaplane tender *Curtiss* and destroyer *Helm* underway and back in action by early 1942.

Difficult as it was to get the less damaged ships underway the hardest task still lay ahead for the sweating salvagers. Sunken and badly holed were a number of capital ships that many thought incapable of being raised much less made seaworthy again. Admiral Nimitz visited the sunken *Nevada* and with new chief Salvage Officer Captain Homer N. Wallin, found the once-proud battleship entirely filled with water, her forward innards completely burned out and her forecastle a twisted shambles from a bomb that had exploded beneath. Nimitz doubted the *Nevada* would ever sail again, but Wallin, one of the Navy's top salvage experts, assured him she would be back in duty despite the shattering damage she'd suffered.

After a massive effort that tested the merit of everyone involved, *Nevada* was refloated by February 12 and dry docked at the now completed Dry Dock #2. It took over 400 dives and 1,500 diving hours to patch up the *Nevada's* badly holed hull; but once refloated and docked work progressed with amazing speed. By late April 1942, *Nevada* joined a homeward-bound convoy for permanent repairs at Bremerton, Washington.

Engineers and experts from the Pacific Bridge Company were of tremendous assistance to the Navy all during the salvage effort. Of special merit was the help they gave in convincing the Navy that cement rather than steel patches could be used to refloat the sunken ships. Using these civilians' expertise greatly accelerated the entire salvage program.

In raising the horribly damaged *West Virginia*, 650 tons of concrete was poured into the jagged holes of the hull before the vessel finally was refloated. The *WeeVee*, as she was affectionately called, had taken seven torpedo hits which virtually laid open her entire port side. Her salvage effort was nothing short of monumental, for poisonous gases, 800,000 gallons of fuel oil and tons of ammunition constantly threatened divers and fitters with the possibility of sudden explosion. Remarkably, the *WeeVee* was refloated without a single casualty and after complete modification and refit at Puget Sound, returned to active duty in time to pump 93 rounds of her 16-inch projectiles back at the Japanese in the Battle of Surigao Straight.

After the Japanese surrender she was the first of the "old battleships" to sail into Tokyo Harbor, a fitting tribute to those who manned and those who saved her from a watery grave at Pearl Harbor.

Few can appreciate the condition under which the salvage crews, especially the divers, performed their tasks. Danger hovered over every move as ammunition, fuel and tangles of wreckage had to be cut away from the bomb-wrecked hulls before pumping efforts could begin. Likewise, work on the burned-out, debris-filled compartments was a tedious duty where a misstep or wrong move could spell instant disaster from explosion, fire or falling wreckage. Yet, to a man everyone engaged in the endless salvage effort worked with a care and dedication seldom seen outside of lifesaving attempts. No matter how impossible the task seemed, civilian and naval personnel cooperated in a grand effort inspired solely by the honest hope of seeing the vanquished fleet get back into action against those who sought to destroy it.

Using every device known to man, and some that hadn't been thought of before, they managed through sweat and skill to get the sunken *California* into dry dock by early April, 1942. She left Pearl Harbor under her own power by October and went on to win seven battle stars against the forces of Japan.

Nor were the capital ships the only ones to receive the tender loving care of devoted salvage crews. The venerable old 4,200-ton minelayer *Oglala* rose from her muddy grave to rejoin the fleet's operation off Leyte and Hollandia. The destroyers *Shaw, Cassin* and *Downs*, battered, burnt-out hulks after the attack, miraculously lived to serve again: the latter two, with only their machinery remaining from the original ship and the *Shaw* with an entirely new hull forward, replacing the one that had been blown off that fateful Sunday morning.

Of the 19 ships sunk and damaged on December 7, only three were never to serve their country again. After a two-year effort to right her capsized hull, the mortally damaged *Oklahoma* was brought into dry dock where a survey showed her to be beyond practical repair. She was decommissioned on 1 December 1944, and sold for scrap in 1946 only to founder in a storm while being towed in mid-1947. The *Utah*, already obsolete when the attack occurred, had been a gunnery target vessel and therefore was of no use to the wartime Navy. She was moved slightly from the position where she sank and served as a diving hulk on which to train new divers.

Last but far from least was the famed *Arizona* which suffered the severest casualties and damage of all. With her midships all but obliterated in the bomb blasts that set off her magazines she was found to be beyond military use. Her topsides were cut away and since the bodies of 1,100 of her crew still remained aboard (including Rear Admiral Isaac C. Kidd) it was decided to make her a memorial to all who fell at Pearl Harbor. *Arizona's* 14-inch guns were given to the Army and a transverse structure was built over the hull on which the memorial and its marble walls were erected.

Today this memorial is all that remains to give mute testimony to the 2,335 servicemen who died on 7 December, 1941.

The war is now only a fading memory, the ships long gone from view, yet the struggle to repair and refit the shattered Pacific Fleet shall remain one of the finest memories of man's ability to overcome a sudden tragedy. The salvages of Pearl Harbor earned their rightful place as true heroes of World War II.

Lost Treasure Of Manila Bay

Colonel Yosho Akido carefully studied the six gaunt American prisoners of war standing stiffly at attention before him. "At ease, gentlemen. Relax. I've some very good news," he said thoughtfully in perfect English. Shifting his glance from man to man, Akido noted their sullenness give way to tinges of curiosity. "Since you are each experienced deep sea divers we find your skills may be of much service to the emperor. If you agree to work with us you will be released from Cabanatuan Prison."

Clad in tattered naval uniforms, the bluejackets silently glanced at each other. The mere thought of being released from their vermin-ridden pest hole was enough to peak interest.

Colonel Akido well knew the deprivations the Americans had suffered since their capture after the fall of Corregidor on May 6. It was now mid-August 1942. Clearly, the anger, resentment and disillusion felt by the American sailors was evident in their weary expressions. "Perhaps you would like an American cigarette?" Akido smiled temptingly, removing a crumpled pack of *Lucky Strikes* from his tunic.

Suspiciously, the divers glanced at each other. Imperial Japanese Army colonels were not known for their kindliness. The colonel must have something very hazardous in mind. Akido handed the pack along with a Zippo lighter to lanky bosun's mate 1/C Paul "Slim" Mann, who in turn passed the much-savored smokes to bosun's mate 1/C Morris "Moe" Solomon. Taking one cigarette,

Solomon sent the pack down the line. "No thanks, don't smoke," muttered bosun's mate 2/C Virgil "Jughead" Sauers, wishing the offering was a cool bottle of beer, not a coffin nail. "But I do," blurted bearded motor machinist mate 1/C Wally "Punchy" Barton, extending his tattooed arm to take the pack from Solomon: quickly lighting up. The last two men in line, CPOs Alan Stanley and Edgar Morse, declined the colonel's offering. Both were hard-bitten Navy chief petty officers wise to the ways of the Oriental mind. Better known as "Boats" to the men, Morse was ranking senior CPO, a career Navy man with 16 years service and the bullish stance to prove it.

Akido's office wreathed in smoke as he continued. "Your work will be humanitarian for it will save many lives," he smiled. "Manila Bay is cluttered with the sunken wrecks of ships lost in the recent fighting. These wrecks are a deadly menace to navigation. As skilled divers you have been chosen to help our victorious Imperial Army clear them. "Any questions?"

"Moe" Solomon cautiously raised his hand. "Ain't salvage work a Navy job, Colonel? What's the Army got to do with it?"

Akido's smile vanished. "Who commands the operation is unimportant. It is a vital task—one to be accomplished with all possible speed. You will be escorted to the train leaving for Manila tomorrow morning. That is all you need know. Thank you. No more questions. Dismissed!"

Soon back in their barb-wired prison compound, the divers quickly set out to find the one American officer who might shed some light on what Akido had up his sleeve. Lieutenant Commander Frank Alfred Davis, USNR, knew everything that was going on inside and out of the camp. During their internment Davis had organized a remarkably effective underground network of Filipinos who smuggled food, medical supplies and information into the prison. Awarded the Navy Cross for heroism as the skipper of the submarine rescue vessel USS *Pigeon* (ASR-6), Davis' opinion was gospel; the divers would trust any advice the much respected officer had to offer. Davis nodded knowingly as the six bosun's mates related the Japanese offer.

"Bullshit: don't believe a word Akido told you. You know damned well they're after the fortune in silver we dumped in Manila Bay. Their intelligence sources are excellent, but they don't know you are the ones who put it there," Davis said guardedly. "Don't let those bastards get their slimy hands on that silver!"

The divers well remembered the backbreaking task they'd performed only a few months earlier, shortly before Corregidor Island surrendered under the weight of Japan's overpowering onslaught. Knowing the end was near, Philippine government officials were determined to evacuate the Philippine national treasury. Several million dollars in gold bullion were safely sent to the United States aboard the submarine USS *Trout*, but almost $9 million (more than 17 million pesos worth 50 cents each) in silver still remained in Corregidor's vaults with no means to save it from capture. The Army quickly devised a plan to dump the silver in the turbulent waters of Caballo Bay off the curled tip of Corregidor where it would be difficult if not impossible to recover.

With dump site coordinates carefully noted by prominent Manila Bay landmarks, the U.S. Navy was given the task of disposing of the silver. This daunting chore fell to Lieutenant Commander George H. Harrison, commander of harbor craft in Mariveles Bay. Rounding up a dozen enlisted men, most of whom were experienced Navy divers who had survived the sinking of the veteran tender USS *Canopus* (AS-9), Harrison then sought the aid of Lieutenant Commander Davis, captain of the heroic repair ship USS *Pigeon*, the largest naval vessel still operating in the besieged waters of Manila Bay. Though hazardous for the *Pigeon* to leave her camouflaged lair, Davis assured Harrison his vessel would undertake the secret mission. Well knowing the dangers involved, Harrison advised the dumping had to be accomplished quickly— and at night — since Corregidor's days were numbered. As it was, the Japanese controlled virtually all of Manila Bay. Few U.S. Navy vessels were still afloat and even fewer dared tread into the range of Japanese artillery able to reach any point in the land-encircled bay.

Despite the risk, in the black of night, the USS *Pigeon* lead two flat-topped silver-laden barges to the drop point with her three-inch guns manned in case the flotilla encountered Japanese patrol boats. Working at feverish pace, the sweating sailors managed to heave the heavy boxes of pesos over the side. Each steel strapped wooden box held 6,000 silver coins; requiring herculean effort to move. Grunting and shoving them into the deep, the men cursed the war, the Japanese and the fact that they faced certain capture. Nevertheless, the job was soon accomplished, completed ahead of schedule; 428 tons of silver pesos had fallen to the floor of Caballo Bay in five wearying nights.

Harrison congratulated his exhausted sailors; admonishing them never to admit to the Japanese where the silver lay. The following morning, May 4, the gallant USS *Pigeon* was attacked by Japanese bombers and sunk in eight minutes. Luckily, her crew was ashore. Two days later, May 6, 1942, Corregidor surrendered. The precious silver had been disposed of without a minute to spare.

Casually gathered around Lieutenant Commander Davis, the men tried not to arouse the attention of the guards as they spoke. Moe Solomon repeated the question he'd asked the colonel. "Why is this a Nip Army show and not a Navy job, skipper?"

Davis frowned, "My guess is the Army wants the glory of retrieving the silver; making a gift of $9 million to the emperor's war chest. But, the silver would be far better spent if it were to wind up in our hands or with the Filipino guerillas."

The men nodded. They understood. Davis was confident the men would do their very best to see that the Japanese recouped little of the taunting treasure lying on the bottom of Manila Bay.

The 90-mile train trip from Cabanatuan to Manila proved a surprise to the six Americans who had endured nothing but poor food and harsh treatment at the hands of their captors. Treated to pork sandwiches and cigarettes by suddenly friendly guards, they laughed among themselves, devising new meaning to the term VIP—Very Important Prisoners!

Bused to the waterfront after reaching Manila, the divers were led to a rusty metal shed overlooking the bay. This would be their temporary quarters until moved onto the diving barge they would call home. After they unpacked their meager belongings, a balding Japanese civilian suddenly appeared in the doorway. Soft-spoken and paunchy, he announced in troubled English that he was a salvage expert with 25 years experience. "I am Mr. Hikobe. Though I am too old to dive we will be working side by side on the 'wrecks.' Come, you must meet the officer-in-charge," Hikobe smiled, gesturing as an Army truck arrived outside.

Captain Horo Takuti, a young officer of impeccable bearing, emerged from the vehicle, bowed graciously and smiled as the men introduced themselves. Takuti invited the Americans to watch his troops unload a tangled harvest of old U.S. Navy diving gear from the truck. Among the array of well-worn equipment were several shallow-water helmets, countless fathoms of curled air hoses, odd tools and several dozen heavy suits of diving underwear. The divers immediately noted that the weighty helmets were the old type long discarded by the Navy because they would fill with water and drown the diver if tilted more than 45 degrees. Worse still, the helmets were designed to withstand pressures only up to 50 feet. Aware, but not admitting, they would be working at a depth over 20 fathoms — 120 feet — the Americans winced. The decrepit helmets could easily collapse and crush a man's skull under such pressure.

The Japanese then walked them to the edge of the dock where a rusting 65-foot barge was tied. Captain Takuti explained in Oxford-accented English that the barge had been the home of Filipino divers assigned to a very special aspect of the Manila Bay salvage program. The Filipinos had been ordered to retrieve a cache of silver coins secretly dumped in Caballo Bay. Inexperienced in working at the depth required, two of the divers had unfortunately died from the "bends." After the agonizing deaths, the other Filipinos refused to dive and promptly were imprisoned.

Only 18 boxes — $55,000 in silver — had been salvaged by the Filipinos. However, despite the poor recovery, Takuti snidely admitted his superiors were confident a great deal more silver remained below. The Imperial Army was most eager to recover the treasure before the Navy learned of its presence, hence, the need to enlist the help of American divers before the Navy intervened and made heroes of themselves in a salvage cache the emperor would be delighted to receive. Listening to the officer, "Boats" Morse smiled to himself. Commander Davis' warning had indeed been accurate.

The truth was out. The greedy enemy was after treasure, not harbor salvage. Informed of the real reason they had been released, the Americans feigned great surprise and confusion over Takuti's revelations. An exceptionally tough task lay before them, the divers insisted. The turbulent waters of Caballo Bay would make recovery exceptionally difficult. Though their captors knew they were seasoned in their skills they were obviously unaware the sailors randomly recruited from Cabanatuan prison were the very devils who had dumped the silver where it lay. This knowledge gave the Yanks an advantage they were not loathe to explore to the fullest. Their Japanese "hosts" had nowhere else to turn except to the perpetrators of the "crime" they were seeking to undo.

"We can't live and work on a filthy scow like that," Chief Morse complained. Bosun's mate 1/C Solomon was quick to add, "Yeah, how do you expect us to do a dangerous salvage job without the right equipment?" Wally Barton, known as "Punchy" to his shipmates, joined the chorus of protest. "We need pleasant, relaxing quarters if you expect us to bust our ass helping you, captain."

Takuti knew they had him over a barrel; this was no circumstance where threats of torture and starvation would intimidate prisoners into accomplishing his aims. "You will get whatever you need," Takuti acquiesced. "My men will help you find everything required. Only hurry, please!"

It would require a week's time before the decrepit barge was cleaned, painted and repaired to the divers' satisfaction. When finished it was the envy of the Japanese soldiers and Filipino workmen who labored around the clock to make it resemble a well-furnished houseboat complete with carpets recovered from General MacArthur's office, drapes, individual staterooms, dockside electricity, plumbing, fresh water, a wood stove, refrigerator and an abundance of fresh food. Having pushed Takuti's patience to the limit, the Americans dared not make any more demands; they were ready to go to work.

Shortly after a hazy dawn a small fishing boat took four of the divers deep into Caballo Bay. With them was Captain Takuti, salvage expert Hikobi, and a new menace in the form of a burly "kempe" wearing the black uniform of the Gestapo-like elite army unit known for its ruthless, incorruptible manner of accomplishing any mission. At a glance, the Americans knew the Kempe with his sumo wrestler-like girth, was no one to fool with.

Within half an hour a rusty barge appeared out of the morning haze. This was their diving platform, Takuti announced. The Americans knew the spot well. The barge was anchored over the exact location of the silver!

Boarding the platform, the kempe took immediate charge. If there was any doubt about their purpose in being on the barge it was quickly dispelled by the sinister giant. "You will recover every box of silver below. You will account for every missing coin, every broken box. And if you lie or try typical Yankee tricks I will have the honor of shooting off your testicles and throwing you to the sharks. Understood?" The Americans understood only too well. The memory of the murderous "Bataan Death March" was still vividly etched in their minds.

The morning's mission was to determine exactly what lay below on the sandy bottom. Vigil "Jughead" Sauers made the first dive. Donning the old helmet they had painstakingly repaired, he descended into the green depth following a manila line anchored to the bottom. The visibility was poor, but even in the dim light Sauers soon saw two mountainous piles of familiar looking boxes towering fifty feet above the sea bed. Almost $9 million in silver lay before him.

From Sauers' very first glance he realized the Japanese had no idea of the fortune beneath their feet, nor the fact that it was conveniently piled in two pyramid-shaped mounds exactly as they had dumped it months earlier. At once he realized too that the kempe was bluffing, for if he knew the silver was so centralized he could easily have Japanese divers recover the entire cache in a few weeks time.

Sauers thought fast and decided to stay below to the limit of his time, then get one of the hauling lines around a sealed box, and signal the Filipino deckhands above to start raising it to the surface via their hand-powered winch. He would make the recovery look as difficult as possible. Biding his time, Sauers swam around the mountainous hoard noticing some boxes had indeed broken open spilling their glittering contents on the sandy bottom. Picking up handfuls of loose coins, he jammed them into his weighted diving boots. Submerged for nearly twenty minutes, he gave the line three tugs indicating it was time to return to the surface.

Looking about as he removed the cumbersome helmet, Sauers saw the Japanese busily examining their trove. Drawing the other divers aside, he quietly told them what he found and what he suggested they do. All of the men agreed. They would make every effort to deny the silver to the enemy. The trick would be to bring up just enough coins to keep the Japanese enticed into continuing the operation. Luckily, working out of sight of the enemy, they would also devise means to set aside as many of the silver pesos as possible for use by Commander Davis at Cabanatuan. The ploy was worth a chance; better than rotting in prison, but if caught it meant their lives.

Moe Solomon was the next to dive. He too brought up one box. Then "Punchy" Barton went into the water, but returned empty-handed. "Dammit, cap'n, the current is crazy down there, first pulling you one way, then yankin' another. The stuff's scattered everywhere. You gotta dig it outta the damned sand with your fingers. My lines keep gettin' fouled and you can hardly see in that murk! It's a bitch, cap'n, sir!" Barton declared convincingly.

The kempe, Hikobe and Captain Takuti intently listened to each report. The recovery would be slow, but they were encouraged by what had been retrieved — the best haul yet, over 11,000 silver pesos! The mission would continue, they announced. Two more dives were made, but no more silver rose to the surface. Last to return as the sun set low, "Slim" Mann dolefully shook his head. "Well, maybe we'll be luckier tomorrow," he offered apologetically. The other Americans winked and grinned.

In the weeks that followed the sailors concocted every means possible to slow the recovery and set aside as much of the silver as they dared for themselves. One hundred twenty-eight feet below the surface of Caballo Bay American ingenuity reached epic resourcefulness. Smuggling tools and marlin spikes to the bottom, the divers managed to break open most boxes being raised to the surface in nets. In this way, the yield would be drastically reduced as the spilled pesos sank to the bay floor. Now loose, the coins were poured into specially sewn bags made from old dungarees attached to well hidden pockets concealed beneath the divers thick underwear. Returning to the barge, the sacks were concealed under tarpaulins and stacks of clothing that would later return with the men on the shuttle boat.

Salvagemaster Hikobi couldn't understand why so many boxes were nearly empty as they where hauled out of the bay.

The divers tried to explain. "Those boxes are rotted. The second you move them they fall apart. Then the current spreads the coins to hell'n gone. We're picking the silver up a piece at a time and putting it in baskets," Punchy Barton shrugged. "Shucks, the war'll be over and we'll still be here suckin' up damn doubloons!"

"You must work faster... recoup more silver," Hikobi angrily urged.

"Dammit, then get us more divers! There's only six of us and lots of that crap still on the bottom," Sauers spoke out.

So far $458,000 worth of silver had been recovered, less than a fraction of what remained below. Hikobi contacted Colonel Akobi at Cabanatuan prison, demanding more American divers be sent.

Two days later three new veteran divers arrived at the dockside barge the American's had dubbed the "Showboat." Carpenter's mate 1/C Harry Anderson greeted old shipmate and fellow New

Yorker Moe Solomon with a welcome smile. With him were Torpedoman 2/C Bob Sheats and bosun's mate 2/C George Chopchick, also friends of Mann and Stanley. After renewing acquaintances the men set about viewing their new quarters. Touring the "Showboat," the grubby-looking newcomers couldn't believe their eyes as Moe Solomon proudly pointed out each unbelievable convenience and comfort. Every cupboard shelf was filled with canned goods, peanuts, coffee, eggs, sugar, cigarettes, rum, nuts, jellies and condiments. "War's hell, ain't it!" Solomon smirked, leading the new trio to a table heaped with fresh fruits of every conceivable variety.

Solomon's expression turned deadly serious. He began explaining the set up, the devious means used to steal much of the silver from under the noses of the Japanese. "This ain't no game, men. We're dead meat if we get caught," Solomon informed them. Each of the new divers were well known to the others. They could be trusted. They'd go along with the program, dangerous as it was.

By then the Americans had diverted more than $200,000 worth of silver from the enemy, a large portion of which had secretly been carried back to Cabanatuan by loyal Filipinos. Once inside the prison it was given to Commander Davis, who used the loot to bribe guards and finance his ever growing spy/support network. But Davis had to be exceptionally cautious less the Japanese learn he had an unending source of scarce silver pesos that could only be coming from one "bank," the squat diving barge in the middle of Manila Bay. A means was soon arranged to channel the silver through money-changers so its wellspring could be kept secret. Giving the needy Filipinos the excess silver allowed them to exchange it through Chinese money-handlers into easily spendable Japanese occupation currency. The Chinese charged a substantial fee to handle the transaction, and the end result was that eventually so much silver went into circulation in Manila that it drastically lowered the exchange rate. After some time the Japanese realized their paper currency was almost worthless, falling 35 to 1, and no one wanted it. Although the Filipinos also took large commissions for themselves, the Americans felt they well deserved every cent for the risk they took with their lives.

Finishing the "Showboat" tour for the newcomers, Solomon opened the secret trap door leading into the bilge. Buried in the slimy water were cleverly contrived pressure chambers from which baskets of stolen silver hung suspended beneath the hull. Conveniently stored in this liquid vault, the silver was easily brought aboard, emptied from the baskets into burlap trash bags, then smuggled into Cabanatuan prison or Manila. Shaking their heads in disbelief, the grinning trio realized they were now an integral part of a vast smuggling operation. Solomon was right, they laughed. War was indeed hell — and profitable!

With nine divers on hand the salvage effort swung into high gear. Each day's recovery doubled that of the day before, greatly pleasing the Japanese, but making concealment of the American's "share" ever more difficult. Bad weather and high seas frequently halted diving, but the men took this time to cultivate new avenues through which to disperse their ill-gotten gains — providing financial support of Filipino guerrilla operations. Inevitably, with so many now involved in the smuggling operation the Japanese were bound to get wind of what was going on.

Early one rainy morning Captain Takuti came aboard the "Showboat" to announce the barge would be moved to the end of the pier next to where the Japanese kept a small tug and patrol boat. Though Takuti's excuse was that this location made it easier to dock the shuttle boat, the sailors suspected the move had deeper implications. Too many Japanese were asking too many questions about the aloof, jaunty Americans who seemed to be enjoying more freedom and a higher standard of living than their captors.

At a round table meeting later that morning the divers discussed the situation. "Moving to the Jap pier means we'll lose our privacy," Boats Morse observed.

Hal Anderson was worried. "Yes... all one of those Nips has to do is swim under "Showboat's" hull and see that forest of silver hanging in baskets! That'll be finito!"

"Word is out all over town. Those damned pesos are showing up everywhere," Punchy Barton added. "Maybe we dump the baskets on the bottom here," Alan Stanley suggested. "That would make it harder to find if they were to pull a sudden inspection." With the Japanese becoming increasingly suspicious, the idea made sense. However, dumping the cache would be postponed by yet another enemy none of the sailors had anticipated — a typhoon!

While discussing their plans the barge began to surge against its moorings. The skies darkened, the rain intensified and the wind soon rose to a howl. By nightfall the storm had grown into a full-fledged typhoon sweeping the entire China Sea. As waves pounded the pier with frothing fury, the Japanese sailors abandoned their wildly pitching tug and patrol boat to seek shelter ashore. But the Americans now found themselves trapped by their own cunning. Powerless, the creaking "Showboat" bucked and rocked, threatening to break the mooring lines and be swept to sea or onto the rocky beach. Laden with stolen silver, there was a distinct danger the barge's fragile wooden hull would split open and spill its illicit cargo on the beaches. If that happened the Americans would pay with their lives, the Japanese then possessing the evidence needed to brand them thieves. With no choice but to save their wave-lashed trove, they secured more lines, locked the hatches and hoped they could ride out the typhoon.

In the typhoon's wake, much of the Philippine shoreline was a shambles. Metal buildings were mangled, boats were tossed inland like strewn toys and few trees still stood. Miraculously, "Showboat" survived.

With murky bay waters forcing a temporary suspension of diving operations, the divers were put to work helping other American POW clean-up parties clear the storm's debris. Working side by side and intermingled with other Cabanatuan prisoners for the following two weeks, Morse's divers had the rare opportunity to pass thousands of the contraband pesos directly to the grateful prisoners of war. But each day saw them more closely watched. Fearing detection, under cover of night, they made good their earlier plan to remove the buckets of silver hanging beneath the barge. Carefully laid on the bottom, the buckets were covered with sheet metal debris.

Hiding the contraband proved providential. The following morning Captain Takuti stormed aboard the "Showboat" followed by three armed guards. "Where is the silver hidden?" he angrily demanded.

"What silver?" Boats Morse asked querulously.

"The silver you thieves have stolen by the ton!"

Moe Solomon became incensed. "Look, cap'n, we've busted our buns for you and now you got the balls to accuse us of being thieves!"

"Yes, that's exactly what I'm accusing every last one of you — wanton thievery...and when we find the silver I'll have you all shot! Search every corner," Takuti shouted to his men. "Showboat" was ransacked to its timbers, the bilge explored and the hull bottom examined by swimmers. The search failed to turn up a single silver peso. Infuriated as his inspection party left empty-handed, Takuti made a parting threat. "We know you have been distributing the silver to the Filipino infidels. The ruse is over. Tomorrow we return to the diving barge to make our own inspection of what lies below." With that, Takuti left.

"It's all over, boys," Slim Mann blurted, watching Takuti drive off. "When they see that friggin' mountain of silver still down there its curtains," Stanley declared.

Bob Sheats wasn't so sure. "What if their diver never comes up again. Suppose he has an 'accident' while he's below?"

"I'd relish cuttin' that SOB's air line," Sauers offered.

Punchy Barton wasn't overjoyed with the idea. "It's no use, gang. They'll blame us anyway. We're dead ducks. What do you think, Boats?"

"One way or the other our little game is over, boys. We better cool it and hope for the best. We've done our damage, helped our buddies, aided the Filipinos and screwed up the Nips economy. Let's call it a day," Morse suggested.

The next day all nine divers were ferried to the diving barge under heavy guard. Captain Takuti made good his threat. Salvagemaster Hikobe accompanied them. The rotund kempe would make the all-telling climatic dive himself. The Americans stood aside quietly watching as the bear-like kempe squirmed into the diver's clothing and helmet. To a man each felt he was attending his own execution, for the unflappable kempe was bound to discover the truth they had kept so well hidden for months — the treasure was enormous in size and easily recoverable.

Holding their breath, helpless to do anything to save themselves, the nine POWs watched as the kempe descended into the sea. Soon the only trace of him was his bubbling air hose. Slim Mann closed his eyes. George Chopchick silently prayed. Moe Solomon quietly hummed a tune. Stanley fingered the religious cross hanging from his neck. The others passively waited.

Suddenly, the kempe broke to the surface wildly flaying his arms, frantically tugging at the helmet as he was quickly hoisted aboard. Barely two fathoms down, the over-sized kempe was afflicted with claustrophobia, unable to continue the descent. Fate had been kind. Their secret was safe. The Americans breathed a welcome sigh of relief. Grinning among themselves, the sailors prepared to begin a normal day's diving.

Unable to prove the guilt of the American divers, the Japanese were content to claim that the silver pesos washed onto the beaches as a result of previous storms and the last typhoon. Eager to save face and protect the integrity of the kempes, the Army vowed the Americans could never have gotten the silver past their watchful overseers. The silver recovery continued into the early Fall, but having saved themselves from embarrassment once, the Japanese Army high command decided it was wise not to provoke any further investigations.

Late in November all diving operations were suspended. The plush "Showboat" was abandoned; the nine American divers sent to Manila to work as stevedores. Ironically, their new "boss" was Lieutenant Commander George C. Harrison, the very officer who had commanded the expedition that dumped the silver in Manila Bay. Like POW Frank Davis, Harrison did everything possible to wage war against the enemy. He encouraged and implemented continuing sabotage of every Japanese ship his POW work battalion boarded. From small holes drilled into freighter bilges to deliberately miscalculated loadings guaranteed to capsize a ship in the slightest storm, "Harrison's 400 Thieves" did their best to disrupt the Japanese war effort.

Of the American sailors involved in the 1942 Japanese silver salvage program all would survive the war except George Chopchick, who died in 1944 aboard a prison ship en route to Japan. Lieutenant Commander Frank A. Davis also died in 1944, still a captive. Later, Davis was posthumously awarded the Legion of Merit for his actions aiding fellow prisoners at Cabanatuan. Davis' USS *Pigeon*, earned two Presidential Unit Citations during her brief, but courageous World War II exploits.

After the war, the U.S. Navy salvaged about $2.5 million of the silver still on the bottom of Manila Bay. Further attempts at recovery by the Philippine government proved more costly than the silver was worth. In the ensuing decades numerous salvage attempts have been made by both private and commercial salvagers, but it is estimated that several millions more in silver still lies on the floor of Manila Bay. Buried and scattered by shifting currents and storms, it may well remain there decades longer as a tribute to the sailors of the U.S. Navy who risked their lives to keep it from falling into enemy hands.

Sea Search

S.S. *Pacific*

Final Voyage Of The S.S. *Pacific*

The Panama Canal had not yet been developed. Most clipper ships making the journey to California took the long arduous route around Cape Horn, finally making port in San Francisco. The trip required months at sea without making port. Fledgling prospectors heard the the shout of "gold" in 1848; ships came to the coast of California from all over the world. The population of California grew from 20,000 to 120,000 in the short span of one year. Excitement ran so high that many of the ships docking at San Francisco wharves had their crewmen desert to head inland to seek gold in the Sierra's.

With all this activity in California, shipping became a very important industry. Many vessels put into service were overcrowded and some ill fated. The more popular ocean-going vessels of the time were American built side-wheel paddle steamers. These ships were often constructed in East Coast shipyards then steamed around Cape Horn to service western ports.

Prospectors, businessmen, and families who preferred travel by boat to California rather than overland stagecoaches could now sail from Boston, New York and Philadelphia taking the new "short-cut route" making an interim landing in Panama.

With the completion of the railroad crossing the isthmus in 1855, more passengers and shippers chose to avoid the hazardous Cape Horn passage under sail for the steady route by the more comfortable steam packets. Upon reaching the Pacific side of the isthmus, passengers reboarded another waiting ship, finishing their journey to San Francisco.

In 1850 the modern side-wheeler S.S. *Pacific* was launched from her ways and designed specifically to transport passengers between Panama and San Francisco. Considered luxurious for the day, the ship remained in service until the Continental Railroad was formed, reducing the east-west journey from months to days.

Loosing most of her revenue and unable to compete with the low cost and now fast moving railroads, the S.S. *Pacific* was placed in the shallows of dry dock in upper San Francisco bay.

Several years passed when another cry of gold was heard! This time, the word came from the Canada and Alaska. Realizing the new potential, the vessel's owners replaced the S.S. *Pacific* into service only now transporting hopeful prospectors from San Francisco into British Columbia and on into Alaska. Eager to get their hands on the gold, prospectors clamored to board the now ageing side-wheeler and were willing to pay the meager passage fee. It was another gold rush promising riches for all.

The 1,200-mile passage from California took five days of sailing along the western coast to reach port in Victoria, British Columbia. In the wheelhouse stood Captain Jefferson Howell, a seasoned master, but a man who had doubts about the *Pacific*'s ability to withstand a foul coastal storm. His ship began showing signs of a dry-rotted hull and balky steam engine. The captain, however, was determined to make port, but assuredly with the first sign of bad weather would take his vessel to the safety of calm water wherever it could be found.

On her first voyage north after being pressed into service, the weather remained in their favor as the *Pacific* continued paddling steadily through rough seas. Just after making port in Victoria, she replaced her stores, unloaded passengers and reloaded her hold with supplies bound for the gold fields. Departure for Alaska was scheduled for the following afternoon.

Now reloaded, the S.S. *Pacific* set sail from Victoria on November 18, 1881. According to the ship's manifest, she carried 275 passengers and crew along with her cargo of food and prospecting tools. Unbeknownst to most, the ship also carried a Wells Fargo safe containing $300,000 in gold. The precious cargo was brought aboard the night before in total darkness and stowed. Jefferson Howell realized its worth and preferred to keep the valuable metal a secret from passengers and crew.

The *Pacific* made way into a cold westerly breeze. While sailing the Strait of Juan de Fuca, the chief engineer complained he had new trouble in keeping the steam required to maintain her 10-knot speed. Captain Howell suggested they return to port for repairs. The engineer explained if he had two hours and could reduce speed to 3-4 knots his crew could mend the problem. The captain agreed, keeping her on course at the reduced speed. If they were unable to solve the problem, however, he would return to Victoria.

Finally, a call from the engine room came informing Howell all repairs were complete, the ship could resume her normal speed of 10 knots.

The ship sailed effortlessly around Cape Flattery. Captain Howell found the wind and waters calming — all appeared well for the *Pacific*. It was now dark; the stars broadcast their beauty in a clear night. Signs of the northern lights flickered above the ship. Pleased they were still making good headway, Howell puffed on his pipe as he stood watch. The ship's position was estimated at 45 miles off Cape Flattery.

Passengers on the vessel had retired for the evening listening to the monotonous sounds of the sloshing paddle wheels turning against the hull.

With his hands held behind his back, Captain Howell paced slowly back and forth peering into the black seas before him. The wheelhouse door flew open permitting cold air to fill the tiny station. Howell spun about; a deck hand rushed in shouting, "Turn to port captain, we're gonna be hit!" With that call, a loud grinding sound was heard and felt throughout the ship.

Howell ran to the starboard wheelhouse door yelling, "What the hell was that?" In the dim light he saw the bow of another ship, the *Orpheus*, backing away from the torn side-wheel of the *Pacific*.

Sea Search

The captain of the *Orpheus* assumed he stopped in time and only brushed the hull *Pacific*. The *Orpheus* signaled she was again underway. In truth, his ship demolished the starboard drive wheel of the *Pacific,* also tearing a gaping hole in her hull just aft the drive shaft. The *Orpheus* disappeared quietly into the blackened night.

Passengers, shaken from their berths by the loud shudder echoing throughout the ship, rushed to the main deck trying to find answers to their questions! "What happened? Are we sinking?"

Still not fully aware of their tragic situation, Captain Howell tried to pacify passengers informing them the ship had collided with another vessel, but there was no immediate cause for alarm. Everything was under control!

Returning to their cabins, acquiescing passengers found others who also were escaping from below. One passenger, dragging his wife by the hand, yelled, "The ship is sinking. Get the hell out of here now! There's no time to lose!" He pointed to water cascading into the companionway behind him.

As the water rose, mayhem spread quickly throughout the ship. Passengers fought their way through dark passageways to upper decks. Fighting over the four lifeboats hanging from davits, deck hands directed women and children to be loaded first. Several men showed cowardice, pushing aside women and children for a seat in the boats.

The ominous ram of the *Orpheus'* bow was more devastating that anyone realized. Another agonizing groan and weird tearing sound suddenly vibrated throughout. The ship mysteriously tore itself in half! Now, separating slowly into two sections, she began sinking into the back icy water leaving 275 survivors to fight for their lives. Those unfortunates who were unable to make the safety of lifeboats struggled in desperation looking for any piece of wreckage to hold on to. Women who failed making it to lifeboats vanished beneath the waves — their heavy clothing quickly absorbed water pulling them under as though tied to heavy weights.

Captain Howell stood outside his wheelhouse looking in horror as the ship disintegrated before him. Although he tried shouting commands, there was so much noise from the yelling and screaming of passengers, that no one heard his orders.

A survivor, Peter M. Bowels, later gave his accounting of that dreadful night:

"I was in my berth getting ready to retire when there was a loud grinding sound coming from someplace toward the rear of the ship. I was sure we hit something. Only seconds seem to pass when someone ran through the companionways pounding on doors shouting, 'Get out! Get out! The ship is sinking!' Immediately, I got dressed and could smell smoke coming from someplace. In the hall, water was already on the floor. Passengers were screaming, fighting to get out. One small child fell and fellow passengers didn't bother to help the child. I have no idea where his parents were. All I remember was grabbing him by the collar and taking him with me up the stairwell to the upper deck.

"On deck, it was very cold. The lifeboats were over filled with passengers. The first boat was already deployed with people hanging on to its gunnels. Flames could be seen coming out of another cabin; I assume a lantern fell causing the fire. The flames did offer partial light and reflections on the water.

"On the second boat, one of the crewmen yelled, 'We're overloaded, it's impossible to raise the boat on its davits, someone must get off!' No one moved, just clinging onto their seats, shivering from the cold.

"With the boy in hand, I ran to still another boat filling fast with grief-stricken passengers. Taking a seat in the front, I grabbed the rope of the block and fall. Another crewmen in the rear of the lifeboat had what appeared to be a fire ax. He too yelled, 'That's it! No more! Okay up there, pull on that line.'

"Now, filled to capacity, we lowered ourselves into the water. It was totally dark. All that was visible were the flames as we cast off. There were only two oars and they were held by women! With so many people on board it was impossible to make headway. Several more people in the water grabbed on to the gunnels, all at the same time! The crewmen stood with the ax in hand, 'Let go or I'll cut your hands off!' he shouted. But his cry was too late; the massive group of swimmers capsized the boat, cascading all of the others into the freezing water.

Men and women were screaming as they fought for their lives. Although very cold, I swam for a large crate bobbing only yards away. I found two women clinging to the box. Both looked at me in the dim light. One lady cried out, 'We can't swim, mister. Please help!' The terrible cold caused them to lose their grip — the two slipped under the water. That crate was my only salvation, I stayed with it all night and finally was rescued by a fishing trawler that took me back to Victoria."

Another survivor of the *Pacific* disaster was a member of the crew, Neilus Hemsley. He too gave testimony of what happened on that fateful night.

"My first recollection as I rested in my hammock was a large crashing sound coming from the side of the hull. My body vibrated and my feet hit the bulkhead from the jolt. Ship's steerage was above us on the between deck. At first I thought it somehow broke loose! My hammock was below on the starboard side. I guess we rested about eye level with the water. Suddenly, seawater came in through the starboard side. Although it was pretty dark, my shipmates and I could see a huge rush of water pouring into the ship. There were no bulkheads between us or the stern of the ship to stop the flow. Water came in through a big hole, it was certainly too big to shore up.

With the water flooding the floor beneath our feet, one of the other men yelled, 'Get out! It's too big to fix, get the hell out!' We scrambled for out lives. Finally, getting to the upper deck, we found people running about in a frenzy! Still others screaming, running in all directions not knowing what to do next! I ran for a lifeboat filled with mostly women. Confused, they had no idea how to operate the block and fall located at each end of the boat. Taking a position in the rear, the man up forward yelled, 'We're overloaded, don't try to lower her. We'll float off when we get to the water.' His call made good sense to me!

"Another man, the purser it think, climbed aboard. I stood alongside of him watching the water rise and the *Pacific* sink before our eyes. There was a huge explosion as the boilers filled with seawater. Thankfully, we started to float off the deck of the ship. The line was still held fast to the fall. With a knife, the purser cut us loose, setting our end free. The man up forward asked for the knife and did the same. We began to float away from the ship, but were tossed about by churning waves caused by the boiler's explosion.

Now, being so crammed with people it was impossible to row. Without warning, a women yelled, 'We're sinking!' The lifeboat apparently became damaged from having been rammed against the ship's hull. We too began filling with water. Passengers tried but found it impossible to bail. The boat flooded completely and sank beneath our feet.

I swam off holding another lady — she too could not swim and slipped from my grasp. Surely her heavy layers of clothing pulled her down. Luckily, I found a severed cabin door, which acted as a float. My body was numb from the cold water. I spent the entire night clinging to it, that door was my lifesaver! A fishing trawler, *Memicis*, picked me up the next morning.

It's estimated, of the 275 persons aboard the *Pacific*, only 33 people survived the initial sinking. Four lifeboats were eventually deployed, two capsized from being overladen with passengers, the third had its hull crushed from ramming into the mother ship and the fourth made it to safety. Of the remaining group, only 25 actually survived the deadly accident. The balance of the passengers and crew met their demise by freezing in the cold water that surrounded them.

Later investigation revealed, that the captain of the *Orpheus* was the one person responsible for the terrible accident. If he would have remained at the site after the ramming, his ship could possibly have offered assistance and saved many lives. In his own defense, he stated he thought there was no damage and gave the command to continue their voyage. "Our ship backed away totally unscathed. There was no reason to go back!" he claimed.

When asked of hearing the distress signal from the *Pacific*'s horn, why did he not return to aid in the call for help?

He commented, "There was too much wind blowing — the horn was not heard from our position!"

The captain was later found guilty of total negligence and leaving the scene of a disaster. The court was to rule on his sentence the following day. Spectators leaving the court wanted to see the maximum sentence placed upon the nonchalant man.

While leaving the courtroom, an unsuspecting assailant drew a gun and shot the captain through the heart. The man died where he stood. The crowd quickly gathered, stunned by the commotion, everyone failed to grab the gunman, permitting him to disappear in the crowd.

It was thought the assailant was an outraged relative of one of the passengers on the S.S. *Pacific* who sought revenge.

To this day, salvagers are still searching for the remains of the *Pacific*. All attempts to retrieve any portion of the vessel or its $300,000 in gold treasure (now thought to be worth an excess of $6 million) have failed.

Treasure Ships
Of New England

All of the New England states have some form of coastline with the exception of Vermont. Because of New England's countless number of coves and inlets, the entire area became a great haven for pirate ships to anchor and hide from their roving pursuers. The natural terrain of rolling hills and tree-lined coves afforded the ships seclusion and privacy.

Names like Edward Teach, alias "Black Beard," Captain William Kidd, John Quelch and Captain Samuel Bellamy, to name a few, have gone down in the history books as the villains of their day. There was so much treasure pilfered from treasure ships returning to European ports, that pirates and their crews were forced to bury their booty in an obscure location. In most cases, pillaging unsuspecting treasure ships became an obsession with the thieves.

Buccaneers already had more loot than they could possibly spend. But, their plight went on. The only temporary solution was to bury their booty. Obviously, their intention was to return later, recover the loot and divide it among the crew. Each was offered his fair share. On most ships, the captain received his two shares, the quartermaster was entitled to one and one half shares and a share to each member of the crew.

Pirate crews were from broad ethnic backgrounds. It was common to find from one third to a half of the men of African background. Most were former slaves who became overseers of their own people and who later turned to the freedom of piracy. Oddly, these men enjoyed an equal share of the booty and were granted the same rights as their white counterparts by the captain.

Few realize there was an on going form of life insurance and a primitive disability paid to crewmen who were injured or killed while serving "in the line of duty." A seamen who lost an eye or limb was granted a handsome additional share of the booty. There were also situations when a man lost his life, often the captain saw to it his family was compensated for the man's death.

Navigation while at sea and map making were limited to a scant few who actually knew or understood the process of working with the compass, sextant and generating defined cartography. The skill of writing down and/or drawing actual records for a recovery was practically nonexistent aboard pirate ships. A majority of the men (and women!) could not read or write and had difficulty utilizing a writing instrument of any sort.

A strict law of the sea dictated that crewmen follow the direct orders supplied by the captain. Questioning his authority or an order, was considered insubordination and immediately punishable — usually, flogging by the lash of the first mate. There were few exceptions.

The majority of the record keeping was performed by the captain and kept in his quarters for safekeeping. Countless loads of pirate treasure were lost due to the looters forgetting where they had buried their booty. Crewmen couldn't decipher or understand even their own maps! The result is that today, there are hundreds of lost or forgotten treasure sites from Maine to Florida.

Typically, during the warm weather months, treasure hunters work the beaches of Block Island, Rhode Island. Modern-day treasure hunters are seeking the buried treasure left by Joe Bradish. His loot was carefully placed in a large chest and hidden somewhere on the northerly tip of the island. Bradish, a pirate, was eventually apprehended and returned to England.

It was said Captain Bradish did make a detailed map showing the exact location of the chest. It was given to a friend for safekeeping. He asked the man to recover the loot and the money to be used to free him from the gallows. The so-called friend was also to testify on behalf of Bradish at his trial. According to the records, the friend and his ship set sail to find the treasure, but were lost at sea in a storm. After a long trial by the admiralty, the pirate was finally introduced to the hangman's noose. The chest is still buried and was never uncovered.

In 1801, an old man by the name of Raphel De los Meo was found digging several large holes just outside a small village that is now known as New Haven, Connecticut.

When questioned by the local townsmen, he told them he was aboard a treasure ship as as young lad. The ship was commanded by a Captain Kidd. The locals were Colonial farmers and tradesman — no one recognized the name of the elusive pirate. Raphel mentioned he was put ashore along with other crewmembers who buried two trunks: one of gold, the other jewels. He explained to the farmers, that the treasure was buried for fear their ship might be overtaken by roaming pirates!

Believing his far-fetched story, De los Meo offered the farmers half the loot if they helped him dig and located the treasure. Immediately, the men went for shovels or any implement that could bore holes in the earth.

The group dug for more than a week with no results. Suddenly, when the farmers were about to give up the search, several gold coins appeared, giving the diggers hope that the treasure was finally located. If nothing else, the find added to the credibility of De los Moe. He was telling the truth — treasure was buried somewhere!

The farmers again took to the search, but were revealed nothing for their effort. The loss was attributed to several large storms that had plagued the area years before. It was possible the chest was crushed somehow and/or washed to another location or, possibly, sunk deeper into the soil. It still awaits detection.

As the hurricane season approached Cuba in the fall of 1752, still having fair weather, the Spanish galleon *San Jose* hurriedly left Havana to make port in Spain. She was carrying a heavy cargo of silver and gold brought from the Spanish-controlled mines of

Central America.

The trade route, favoring Havana to Europe, was to sail north along the Florida coastline running with the gulfstream then turn east when abeam the Carolinas.

Prior to the *San Jose* coming about for her easterly track, a seaman cried out to the captain that the ship was taking on water in the forward cargo hold.

The first mate was ordered to investigate the trouble. The leak was detected in the bow along the keel. The mate reported the severity of the trouble to his superior. Feeling he may need help, the captain displayed a distress flag notifying passing ships of her situation.

The American ship *Susannah* came alongside offering help. The seaman finally got the problem under control and the ship could go on, but the repairs were no more than a temporary fix.

The captain of the *Susanna* saw the problem and felt it may become more serious than even what was reported. He questioned whether or not the ship was capable of making the journey to Spain especially during the threat of the hurricane season. It was suggested the *San Jose* follow the *Susannah* for repairs in New London, Connecticut. There, they had the proper facilities to repair the leak.

Recognizing this as a wise move, the Spanish captain agreed lumbering along slowly behind the *Susannah* into port.

Not being familiar with the New London harbor, in error, the *San Jose* ran aground while lowering her sails. Due to extensive weight, she was forced to unload all of her ship's stores including the gold and silver onto the *Susannah*. Hopefully, this would free the ship and she could be towed to a nearby wharf.

Thinking the 50 treasure chests laden with silver cobs and gold doubloons were actually stolen by the officers and men of the *San Jose*, the local government seized everything until the rightful ownership could be proven. The chests were transferred to an undisclosed warehouse.

Outraged, the Spanish captain put up a protest to the government! A formal hearing was initiated — finally he proved his innocence and that he and Spain were the rightful owners of the treasure. The court finally agreed to return the chests to the galleon and let the Spaniards continue their journey to their homeland.

Repairs to the *San Jose* were finally completed. The chocks were pulled and the ship slid down the ways. The *San Jose* was again ready for sea.

Satisfied with the outcome of the trial, the captain patiently awaited for the delivery of the treasure from the warehouse. He stood on deck and was most anxious to get underway to catch the next high tide. Four of his men appeared running along the dock. They carried a single chest. Out of the 50 chests that were to be delivered to his ship, all but one were missing. The balance of his valuable cargo had mysteriously disappeared!

His rage again grew. Another protest was initiated with local authorities. An immediate search was put into motion, but ended in a futile attempt to locate any sign of the treasure.

It was thought the unknown thieves ran off with the chests and due to the large volume of coinage, they no doubt had to bury the money within the proximity of the warehouse. Somewhere it still remains!

In 1720, a ship-rigger by the name of Ned Low decided to turn pirate and give up life as a landlubber. For most of his early youth he was a petty thief and street bully. When Low finally took to the sea out of Boston harbor, he was considered a tough sailor even by pirate standards! Low only knew one swift way to get the truth from his unfortunate victims — torture! It worked every time.

Off the Carolinas, Low and his band of men spotted a Portuguese ship riding low in the water. He visioned her holds were heavily laden with treasure or tradable goods. Peacefully, he called upon the Portuguese captain asking for his help. He cried out that his water casks had turned rancid and they were in need of a fresh water supply.

The captain obliged, pulling alongside. Low boarded the unsuspecting ship in a fury of rage. It was only minutes before they took complete possession of the vessel. Low called for a thorough search for the ship's treasure. He asked members of the crew for the exact location. Bitter by the ferocious boarding party, the men refused to talk. Nothing was found. Angered by the lack of information, Ned Low had the unfortunate captain tied to a mast.

He personally placed the edge of his sword against the man's throat. The captain still would not reveal where he had hidden the ship's treasure. With a fast swipe of his cutlass, Low slashed off the man's ears and lips.

Ned Low gave the order for each severed piece of the man's anatomy to be boiled in water before members of the crew. He then ordered the ship's first mate to eat them! Shocked by the order, the mate quickly pulled a locker key from the captain's waist and handed it to Low. The mate then led his capturer to a secret locker in the captain's quarters.

Captain Low retrieved almost 20,000 moidores from the locker. Satisfied, Low then loaded his ship with the loot and balance of the ship's stores. In celebration, his men murdered the entire Portuguese crew one by one, tossing their remains to the sharks. As a parting gesture they set fire to the abandoned vessel, thus eliminating all traces of its whereabouts.

Fearing they may be caught with the stolen booty, Captain Low set sail for the New England coast. He landed on what was thought to be Lovell Island and had the moidores buried by two of his crew — Low merely looked on.

When he returned to his ship, Low was alone. The crew immediately became suspicious. He told fellow deckhands, the two men with him who helped bury the treasure fled off into the woods. "They were mutineers." he cried out! They accepted his word, but in truth, all hands presumed Low killed the men since they were the only others who knew where the loot was buried.

It was assumed Low had made plans to recover the booty several months later when he had a dispute with his quartermaster. To his dismay, the man took sides against him. He was overpowered and sent adrift in a longboat without food or water.

In 1724, the French caught up with him and several of his crew.

Treasure Ships Of New England

By this time, Low had gained a reputation of being a vicious pirate and possible madman. His acts of cruelty were known by all of those who sailed the sea. With their identity confirmed, each pirate was sent to the gallows. Because of his well-known brutality, the hangman saw to it he died in a slow torturous manner.

The treasure on Lovell Island was never recovered and is thought to be located on the northeast, lee side of the island.

There was an accounting that the famous Captain Kidd had a large hoard of treasure aboard his ship. He and his men fought their way from the warm Caribbean waters overtaking precious treasure ships bound for Spain. With their belly full of gold and jewels he turned north.

Wanting to elude being caught by larger warships of the line, he opted to hide in the still backwaters of Newport, Rhode Island. Kidd ordered the long boats to be lowered to replenish the ship's stores. At the same time, the precious booty was to be brought ashore and buried for safekeeping.

The chests were no sooner lowered into the earth when a large group of Indians raided their site. Several of his men were killed instantly. Realizing his party was outnumbered, Kidd ordered a retreat for the balance of his crewmen to flee back to the longboats and finally the ship itself for safety.

Kidd sat and waited offshore for the Indians to disburse. They too had patience realizing they had his needed food supply. The captain was no fool. His odds of winning a fight were slim. Kidd weighed anchor and sought to find another location farther south for his needed supplies. Captain Kidd thought it best to leave the area and return to the Caribbean in search of new prey. The chests were never reclaimed.

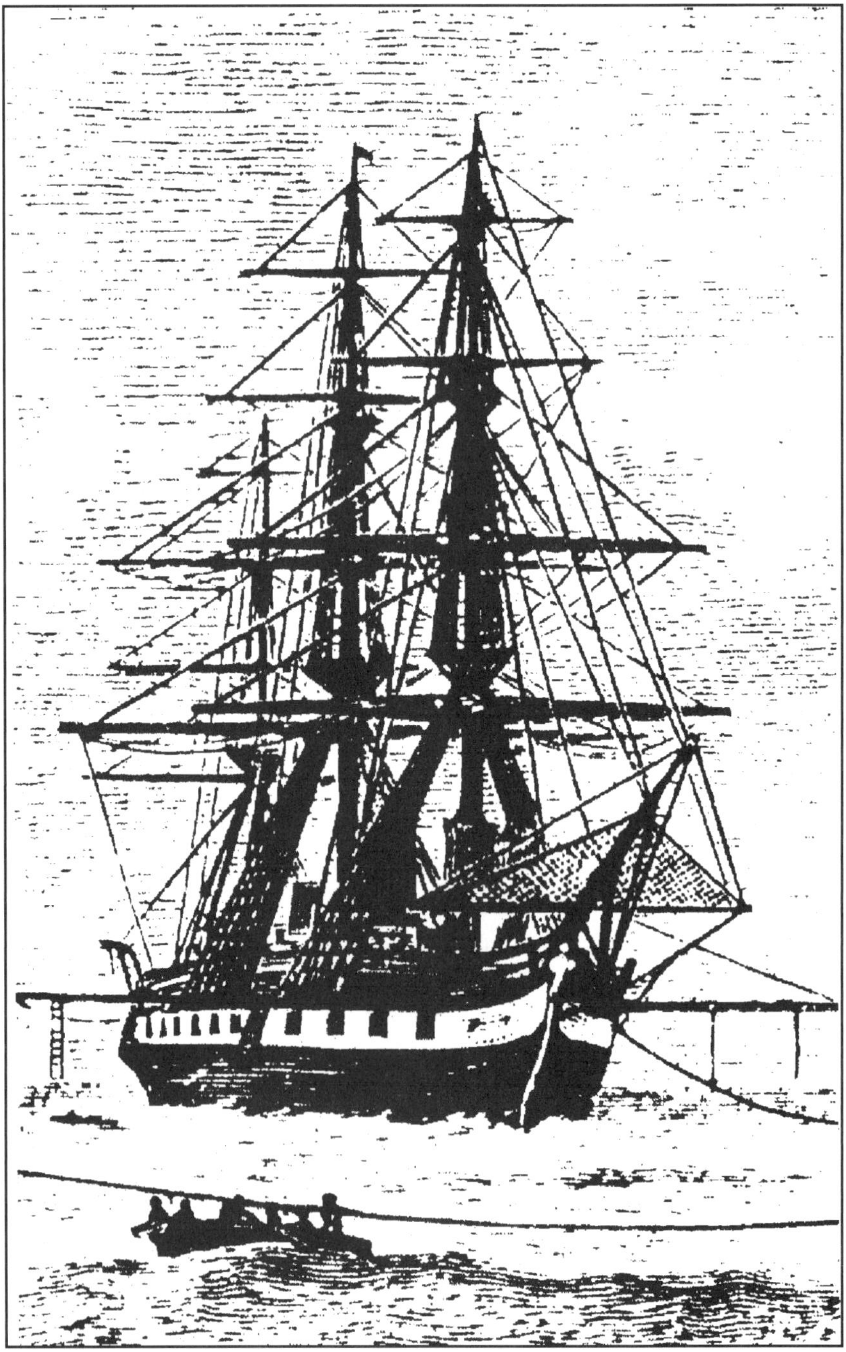

The Big Boom In Treasure Salvage

Inspired by the findings of many lost ships like the *Bismark, Atocha* and *Titanic*, treasure hunters are using modern technology to probe deep for the riches of the sea. But with record finds making daily headlines, a whole host of new questions about ethics and law is being raised...

Early in June 1989, a group of California divers made a pilgrimage to the murky ocean floor off Anacapa Island to explore the wreckage of a long-sunken gold ship, the paddle steamer *Winfred Scott*. Sport divers all, the group was unique in that they were not treasure hunters responding to the lure of lost gold, but an organized class of amateur archaeologists learning how to study and record the rich lore of sunken ships for preservation rather than plunder. As they descended to the depths each diver well knew what he could expect to find, for the tragedy of the encrusted wreckage beneath them had been astutely studied in a classroom under the auspices of trained professional divers and marine archaeologists.

It was on December 2, 1853, that the *Winfield Scott* went down. Just one day out of San Francisco, Captain Simon F. Blunt was attempting to save time off his run to Panama by navigating between the Channel Islands, rather than seaward of them. Suddenly, scores of sleeping passengers were thrown out of their bunks as the steamship ran full speed into an outcropping of rock in a heavy fog.

"Our predicament seemed awful in the extreme," passenger Edward Bosui wrote in his memoirs, noting that the ship lay stuck

Sea Search

on the rocks amid the crashing surf.

Within hours, the *Winfield Scott,* a 225-foot wooden steamer that regularly ferried Gold Rush passengers up and down the coast — had settled on the bottom with $801,874 in gold and tons of mail, cargo and baggage in her hold. Miraculously, all of the passengers survived, spending several days camped out on the barren island before being rescued by passing vessels.

The recent visitors to the site of the wreck were students from California State Long Beach. Experienced scuba divers, they were members of a hands-on course in underwater archaeology that is the first of its kind in Southern California. The course is sponsored jointly by the university and the Los Angeles Maritime Museum. Its mission, among other things: to provide an antidote to private plundering and bolster a national trend by helping state and federal archaeologists map and document shipwrecks such as the one off Anacapa.

"They're helping us with our work," said James Delgado, maritime historian for the National Park Service. About 14 miles out of Ventura, the wreckage off Anacapa is in waters controlled by Channel Islands National Park.

"This is a historically important wreck and the students are adding to our understanding of it." Delgado said. Hoping to dive into the waters of the early morning, the 21 students had left Ventura aboard the California State research vessel *Yellowfin*. They were accompanied by Donald P. Morris, a park archaeologist. Their immediate goal was to expand and clarify the preliminary, partial site map they carried, which park rangers started to draw in 1983. Descending 30 feet to the wreckage, class members held crude plasticity maps while fanning the sand away from encrusted pieces of debris in search of artifacts.

To the untrained eye, the *Winfield Scott* no longer much resembles a ship. After 135 years underwater, the once-proud vessel now consists in the main of a series of encrusted mounds encasing debris, scattered over several hundred yards.

Most of the gold that the ship carried has long since been picked up by scavengers. What remains of the vessel is a large, circular structure believed to be part of the paddle wheel; several bits of iron machinery, including the boiler, and a large number of artifacts from shipboard stores, passengers' baggage and provisions. Hidden in the sand, the objects are sometimes uncovered by the underwater swirl of storms or the fanning of divers.

The first dive went uneventfully. Working in teams, the divers spread out over the wreck to survey its extent in smooth, clear waters. Then the sea kicked up, making the second dive difficult. Using all their strength, the divers swam against a current to reach the wreck and sift through its sand as the murkiness engulfed them.

It was a task for which they were well-prepared. Meeting weekly for 10 weeks, the students had heard presentations from experts on topics ranging from how to survey and map an underwater archaeological site to how to collect and preserve artifacts.

Offered through the university's extension program, the course carried three units of credit. In addition to attending lectures, the students — representing a variety of disciplines and professions — were required to write papers on various California wrecks, relying on the original research.

To be sure, they were not the first to visit the watery gravesite of the *Winfred Scott*. They were, however, the first to visit it for a serious scientific purpose.

That purpose was enhanced last year when federal lawmakers enacted legislation called the Abandoned Shipwreck Act.

The law is meant to protect the 50,000 to 100,000 shipwrecks — including several thousand off the California coast — lying in waters regulated by the state and federal governments. For the last 200 years, they have been governed by federal admiralty law, which asserts that whoever finds a wreck has the right to salvage it and keep at least some of the take. While some states had enacted their own shipwreck legislation, the courts had been inconsistent in declaring whether state laws took precedence over admiralty law.

In 1988 legislation changed all that. First it assigned ownership of historically significant shipwrecks to the federal government, which then transferred title to the state in which they were found. Second, it encouraged each state to establish a program to study and protect its wrecks.

In California, many see the California State Long Beach course as the beginning of just such a program.

"There are a lot of ships out there deteriorating and there's not enough money or trained archaeologists to study them," said Jack Hunter, a professional marine archaeologist who taught the course with William B. Lee, director of the maritime museum in San Pedro. "What we'd like to develop is a cadre of knowledgeable (amateur) divers around the museum to actively survey and investigate shipwrecks with scientific precision and expertise."

Added Lee, who said the museum plans to offer the course regularly: "There's an endless supply of volunteer sport divers who eventually get tired of just going out for game. We can recruit them for this much-needed work." Until recently, it seems, many of those divers — who number an estimated 56,000 in California — were part of the problem. Equipped with private or chartered boats, they routinely raided coastal shipwrecks, retrieving the choicest artifacts as mantel decorations. A few years ago, 19 divers on a single boat were cited for illegally removing artifacts from the *Winfield Scott*, which, because it is in a national park, receives special protection.

The incident sent shock waves throughout the sport-diving community, with most of those cited eventually paying heavy fines.

Archaeologists say that such plundering is extremely damaging to their efforts to reconstruct the past. Once an artifact is removed without documentation, they say, much of its value to history is lost. This is especially true in a shipwreck such as the *Winfred Scott*, which affords one of the few remaining examples of the maritime history of the California Gold Rush, a particularly colorful and important period in the state's past.

Recently the ship was deemed of sufficient historic interest to be placed on the National Register, a list of historically significant sites maintained by the National Park Service. Only five other

The Big Boom In Treasure Salvage

California shipwrecks have rated that status, according to Delgado, with another eleven nominated and awaiting approval by the Secretary of the Interior.

The students found nothing of significance on their recent Saturday dive. But they vowed to return, and Morris said they will be welcome.

"It's slow, painstaking work," he said, "but I think it represents a better attitude than in the past. We, as a society, are realizing that there are valuable lessons to be learned underwater, that these wrecks are good for more than producing souvenirs."

Much of the reason for the present federal legislation regarding sunken ships is the result of the big boom in treasure salvage that has taken place in recent years. Thanks largely to new technology in locating sunken ships, treasure salvage has become big business. Although few of the treasure hunters enjoy the sophisticated deep-diving submarines such as those used by Dr. Robert Ballard to find and explore lost wrecks like the liner *Titanic* or the more recent German battleship *Bismarck*, the fact remains that major treasure finds have been making headlines for the better part of 50 years and the end is nowhere in sight. Even the new federally mandated restrictions on the plundering of lost ships seem to be little more than just another bureaucratic implement to treasure hunters determined to find their pot of gold in the murky depths of the sea.

The late treasure hunter Mel Fisher, who found the gold-laden *Atocha* in 1985, claimed he located the wreck of another Spanish galleon said to contain more gold than the prize Manila galleons. In San Francisco, divers have recovered some of the $2 million in gold and silver from the liner *City of Rio de Janeiro*, sunk in 1901 outside the Golden Gate.

Some 160 miles off the South Carolina coast salvagers announced have found the wreckage of the gold-laden steamboat *Central America*, sunk in a hurricane in 1857 with $450 million in fresh California gold bars aboard.

Are all of these finds in recent years a mere coincidence, or has there begun a sudden resurgence in deep-sea treasure hunting? The answer is simple and has nothing to do with coincidence. Treasure salvaging has become big business, so big that venture capitalists are threatening to make this kind of risky enterprise the "in" investment in the decade. It's all highly speculative and most expensive with risks that border on the ridiculous, but when the stakes run into the multi millions the lure of an enormous return has great appeal to stouthearted investors.

Why is treasure salvaging suddenly in the limelight? The answer is modern technology. Today, as never before, shipwreck salvaging has the benefit of state-of-the-art sonar technology, deep-sea robot explorers and a unique blend of other sciences and skills that didn't exist a decade ago. While numerous other finds like the discovery of the *Lusitania* made brief headlines only years ago, it was the dramatic photos of the *Titanic* and Mel Fisher's record $400 million recovery from the *Atocha* that brought treasure salvaging into the international spotlight.

"Interest in shipwrecks has never been higher," said Ross Holland, president of the National Foundation for Maritime Conservation. "The lure of finding vast caches of sunken treasure has captivated man's imagination for centuries. But in the past, treasure hunting has been a hit-and-miss affair with, I'm afraid, more misses than hits. Today, that has all changed, for with modern technology and astute research amazing recoveries can be made."

With interest fueled with powerful computers that correlate historical data such as wind speeds, tides, ocean currents and sailing times, treasure hunters now can reasonably pinpoint wreck sites with a degree of confidence never before possible. The discovery of the *City of Rio de Janeiro* emphasizes this, for though the liner sank in San Francisco Bay within earshot of a lighthouse near Point Lobos, deep currents drew the hulk far from the rocks the liner hit in 1901. The loss of the passengers made the tragedy the worst maritime disaster in San Francisco's history. Divers who sought the wreck were amazed that no trace of the vessel could be found anywhere near the spot where it sank, almost in view of the rocky shoreline. The wreckage has eluded treasure salvagers looking for the ship's $2 million in silver bullion for more than eight decades. The group of explorers who pooled their resources to attempt the search methodically reviewed every known reference and record about the liner's sinking and then put it all together on a computer.

High-tech treasure hunting isn't easy, nor is it inexpensive. The tools of the trade are chiefly new sonar devices capable of searching three-mile-wide swaths of the ocean floor in one pass, a vast improvement over sonars that could cover only a few hundred feet a few years ago. While the sonar makes its broad sweep of the bottom, computers are processing information already stored to produce a vivid, easily decipherable mosaic picture.

When the wreck is found the real work begins, with sophisticated, deep-diving remote-controlled devices to explore and probe the wreckage. It is these articulated miniature un-manned submersibles that transmit the amazing pictures and retrieve the rich artifacts like the jewel-laden leather valise from the *Titanic*. Capable of exploring the murky passageways of sunken liners at almost any depth, these remarkable robots search and delve far deeper into any wreckage than is safe for a human diver. Above, aboard the salvage ship, human eyes watch TV monitors seeing what the robots finds with relative ease and safety. No life is in jeopardy and the only item at risk is the robot itself — alone and often worth more than a million dollars.

All of these vastly expensive, high-tech commodities came together in the search for the steamer *Central America*. Wreck discoverer Thomas Thompson calls it "one of the greatest shipwrecks on record prior to the *Titanic*."

The broken wreck of the *Central America*, a U.S. Mail steamship that went down in a hurricane with the loss of 428 lives, lies in 8,000 feet of water just within the 200-mile continental limit off the South Carolina coast, says the Columbus America Discovery Group.

The limited partnership based in Columbus, Ohio, found the wreck during the summer of 1988. It made public its discovery after suing successfully the U.S. District Court in Norfolk, Virginia, for jurisdiction over its claim.

The sinking of the *Central America* was nearly as famous in the 19th century as that of the Titanic in the 20th century.

The golden age of American sail, which began with the fast clipper ships of 1848, reached its apogee in the Gold Rush years as the square-riggers raced around Cape Horn from New York to San Francisco in as little as 90 days.

But completion of the railroad across Panama in 1855 spelled the death of the clipper ships. More and more passengers and shippers chose to avoid the hazardous Cape Horn passage under sail for the steady route by steam packet down the coast to the isthmus and across by rail.

The *Central America* had completed 43 such voyages between New York and Panama when it left the isthmus in September 1857, carrying a regular monthly shipment of gold from the new San Francisco mint to the banks of New York.

Some historians say its failure to arrive prevented the banks from staving off the Panic of 1857, which was one of the major economic depressions of the 19th century.

The gold was a government shipment valued at $1,219,189 in 1857 dollars, when gold was valued at roughly 90 cents an ounce.

Records show 103 crew members and 478 passengers — many of them homebound prospectors carrying their own gold — were aboard the *Central America* when it left Havana headed north for New York. Three days later it steamed into a hurricane and began to take on water.

As the rising water extinguished the boiler fires and power to the pumps, the passengers bailed with buckets for 30 hours, buying enough time for all 31 women and 29 children, and 39 of the male passengers and crewmen, to be safely transferred by boat to a nearby sailing vessel.

But the steamer sank shortly after dark September 12. Fifty-four men were later rescued from the water by other passing ships, but 428 passengers and crewmen were lost.

Thompson, a mechanical engineer from Ohio State University

The Big Boom In Treasure Salvage

who had worked as a research scientist to ocean mining projects, said he first learned of the disaster in 1977 after researching the shipwrecks as a hobby.

He became convinced that accelerating advances in undersea technology would make recovery of deep-ocean wrecks possible some time in the 1980s.

The catalytic event, he said, was the development in 1985 of a computer-based imaging system that permits ocean searchers to view on a video screen the profile of objects detected by a towed apparatus that bounces sound waves off the ocean floor.

"I got interested in deep-water ocean wrecks like the *Central America* because they were less likely to have been disturbed by other salvagers," he said. "But there was no way to do this then."

Thompson was particularly interested in the *Central America* because of its unique side-wheel architecture. "They were only made for about 30 years and only nine sank. So, if we find something that looks like this, we have a good idea what it is."

By July 1985, he was convinced that search technology had ripened. Taking a leave of absence from Battelle, Thompson formed Columbus-America Discovery Group and raised $5 million from local investors for his search.

Operating out of a large Victorian house in his native Columbus, he assembled a staff of 11 historians and researchers to gather everything known about the *Central America* and the circumstances surrounding its sinking.

The data was then correlated in a computer relying on techniques first used to find an H-bomb that fell harmlessly into the Mediterranean in 1966 when a B-52 bomber and a KC-135 air tanker collided.

The computer narrowed the search to a 1,400-square-mile area on what is called Blake Ridge, 150 miles off the South Carolina coast.

"The ridge is virtually ideal for sonar searching because it is flat and featureless, and anything like a ship will show up clearly," Thompson said.

But the most useful tool of all came with the new adaptations in the use of side-scan sonars.

In the side-scan sonar, a "fish" towed behind a ship emits a narrow band of sound waves toward the ocean floor.

Sensitive microphones pick up the echoes that bounce off the objects or the sea floor, and the computers use the elapsed time for each echo to calculate the distance to that point.

Side-scan sonar was developed in the 1970s to help find manganese nodules on the ocean floor; recovery of these nodules did not prove economically feasible, however.

During the 1980s, companies like International Submarine Technology of Seattle refined the technology to give much better pictures of the ocean floor. In particular, the company improved it so that the device could cover a swath 15,000 feet wide rather than the several hundred feet covered by earlier models.

The sonar units cost anywhere from $300,000 to $1.5 million. "But what we are interested in was the reduction of risk rather than the reduction of cost," Thompson said. The actual search for the *Central America* took 40 days.

In 1985, the search for the *Titanic* also had taken about 40 days. But that search covered only 140 square miles less than a 10th of the area searched by Thompson. About 80 percent of the *Titanic* search was performed with less sophisticated side-scan sonar owned by the Paris-based oceanographic research institute IFREMER. The ship was actually located during a search with television cameras by Robert Ballard of the Woods Hole Oceanographic Institution in Massachusetts.

"During the search for the *Central America*," Thompson said, "the hunters found about 400 'anomalies,' nine of which were later found to be wrecks. Most of the rest were 55-gallon drums that were probably discarded from ships."

The group kept its discoveries a secret and went back in May to resume surveying the site. They were forced to go public when a second group of searchers appeared on the horizon.

The second team was financed by a group of investors called Boston Salvage Consultants, Inc. A partner in the group, Boston investment banker John O'Brian, refused to discuss details of the operation, but said that his group's financial resources were "at least as great" as those of Columbus-America.

The Boston group had hired the services of treasure hunter Burt Webber, who in 1979 found the Spanish galleon *Concepcion* in a coral reef 85 miles north of the Dominion Republic. Webber in turn hired William Ryan of Columbia University's Lamont-Doherty Geological Observatory in New York City. Ryan, who had searched unsuccessfully for the *Titanic*, has sonar apparatus similar to that used by Thompson.

The third group, three businessmen from Georgetown, South Carolina, incorporated as the South Carolina Marine Archaeological Trust, had hired Steadfast Engineering of Norfolk, Virginia, to find the *Central America*. Steadfast frequently had been hired to locate the remains of plane crashes in the ocean.

Alarmed by the latecomers, Thompson's group quickly recovered wood, coal, and metal samples from the debris surrounding what they believed to be the *Central America* and took them to the U.S. District Court in Norfolk.

On July 16, Judge Richard B. Kellam appointed Columbus-America custodian of the wreck and issued an injunction barring all others from entering a 15 square-mile area around the wreck.

Kellam claimed jurisdiction over the wreck, even though it is outside the U.S. territorial waters, because the vessel was an American ship, because the discovering group was American and, most important, because the first salvage remains were brought to an American port.

Kellam's ruling was "unprecedented — a novel exercise in admiralty jurisdiction," said San Francisco maritime law expert Graydon Sterling who was not involved in the case. Boston Salvage appealed, claiming that Kellum did not have the jurisdiction, but the U.S. 4th Circuit Court of Appeals refused to stay the injunction while the appeal is pending.

Kellum's was the second critical legal decision made affecting treasure hunters. In May, a Massachusetts Superior Court ruled that the "finders keepers" federal admiralty laws have precedence over the rights of individual states and that the states cannot confiscate a share of booty from wrecks from within the three-mile territorial limit.

Both decisions should have the effect of further stimulating treasure hunting because they increased the likelihood that a treasure seeker can retain salvage rights, maritime layers said.

Meanwhile Thompson's group continued to survey the *Central America* site, producing stereo pictures, videotapes and photomosaics. They are also collecting artifacts in small amounts from the debris. This work is accomplished with specially designed equipment that is suspended from the surface, unlike the manned submersible Alvin used to explore the *Titanic*.

Because the equipment is suspended, it does not require vertical thrusters — "so you don't take a risk of destroying artifacts and that visibility is much better," Thompsom said. "And because it is operated from the surface, it can work 24 hours a day."

As of this date several million dollars has been spent on the project. The treasure that has been brought up has unconceivable value.

But scientists fear that they will lose information anyway because there are no archaeologists involved in the operation. "Once you remove artifacts from their context, their scientific value is destroyed," said marine archaeologist Barto Arnold of the Texas Antiquities Commission in Austin. "It's the relative location that tells the story."

Some scientists fear a repetition of the events that occurred during the salvage of the DeBraak in August of 1986, which underwater archaeologist Allen Albright of South Carolina Institute of Archaeology in Columbia, South Carolina, calls "a horror case." Salvagers hired by New Hampshire developer L. John Davidson has tried to raise the 18th-century British warship from the Delaware Bay before the artifacts were removed. During the process the ship tilted and everything in it was dumped onto the bottom where much if it was lost.

The Big Boom In Treasure Salvage

"Commercial treasure ship salvage is just not compatible with archaeology," Arnold said.

The shipwreck finders, of course, do not consider themselves looters. "We have a responsibility to preserve the history and the culture and those things that have to do with the shipwreck itself," said Thompson, who argues that his group did a good job documenting the *Central America* site.

"All of us are interested in saving historical vessels of interest to the American public," said marine archaeologist John Broadwater of the Virginia Department of Conservation and Historic Resources in Yorktown. "We're fervently hoping it will eventually get resolved."

Meanwhile, the big boom in treasure salvage goes on yielding rich rewards for the hunters. The riches may not always be gold or jewels, but the historical and cultural worth of these artifacts are inestimable. Divers sponsored by the Australian Bicentennial Commission early in 1987 recovered hundreds of valuable artifacts from the *Sirius*, flagship of the British fleet that transported more than 1,000 convicts to Australia in 1789. Booty taken from HMS *Bounty* has been taken from the wreck of the *Pandora* which was carrying some of the mutinous crew when it sank off the northern end of Australia in 1790.

The famed Confederate ironclad lies upside-down in 225 feet of water 16 miles off the coast and is rapidly deteriorating according to divers who have been on the wreck. Efforts will be made to preserve this historically invaluable warship that gained such notoriety in the Civil War.

But while the hunt for invaluable history motivates many, it is the lure of golden riches that motivates most. Treasure hunter Bert Clifford continues to recover treasure from the slave ship *Whydah* off Cape Cod, Massachusetts. Seized by the pirate Sam Bellamy in 1716, the *Whydah* ran aground in 1717 taking with it all of the fabled pirate's loot and treasure stolen from countless other ships. Clifford's team has already removed $25 million in gold coins, jewels and artifacts.

Farther south, off Nantucket Island, treasure salvor Bill Fowler's cadre of divers is bringing up gold and treasure from the luxury liner *Republic*, sunk in 1909 after it collided with the Italian liner *Florida*. Quite possibly the *Republic* may yield the all-time high for recovered wealth for it was carrying well over $3 million in gold coins purchased by the Bank of France and is now calculated to be worth as much as $1.7 billion. On the West Coast treasure hunter Robert P. Mester commenced salvage efforts on the liner *Governor,* struck by the freighter *West Hartland* in Admiralty Inlet near Puget Sound in 1921. Sunk in 220 feet of water that features dangerous tides and currents, the *Governor* is estimated to contain nearly $9 million in cash and jewels in her rusting hull. Efforts to bring the liner's safe, reportedly containing $1.5 million in cash alone, have been in process for some time.

The search for gold — as timeless as time itself. The lure and the promise of sudden riches still beckons many — only today it is becoming big business for all involved.

Hey Look! It's The *Great Eastern*

The *Clermont* made her initial run up the Hudson River to Albany, New York, in August 1807. Designed by a civil engineer from Pennsylvania, Robert Fulton, it was the first successful attempt to propel a ship by means of steam power, a Boulton and Watt steam engine. The river run, called "Fulton's Folly" considered a feeble attempt to drive a ship without the use of sail. Witnesses who stood along the banks of the Hudson River laughed as the 150-foot *Clermont* passed with her engine chugging and smoke spuming from her funnel. The little ship's forward speed was only four-knots, but nonetheless, she defied the laws of the sea and moved without the use of wind. On her first voyage, the Clermont's 15-foot diameter paddle wheels took her from New York to Albany and back, a distance of 150 miles, in an astonishing 62 hours.

Despite the controversy of her unorthodox design, the *Clermont* continued servicing the river for almost eight years after that eventful day. Fulton's concept became the prototype of several ships that followed. Soon thereafter, small shipbuilders began developing steam-powered engines that were applied to riverboats and the original tugboat concept. Few marine designers of the day, however, thought the concept would ever be used for trans-ocean crossings. The engine was new, difficult to operate and unreliable. The paddle-wheeled ships that followed Fulton's prototype were capable of both speed and maneuverability and were well suited to river and bay transport. Paddle wheels, however, were a disadvantage in the rough waters of the oceans.

In 1818, the 110-foot U.S. Ship *Savannah* was sent down the ways in New York harbor. She was different than the other ships tied to neighboring piers. The *Savannah* too was powered by an auxiliary steam engine and wielded a set of paddle wheels that had the ability of being folded and stowed when not in use. In her hold she carried nearly 60-tons of coal to stoke her boilers. The engine, a single piston of ultra-simple design, was used as a support device to aid her gleaming suite of sails.

There was still much in the way of opposition by large ship owners when it came to steam power aboard their ships. Potential investors commented, "Why use foolish steam when the wind is free? Look at the danger of having a continuous burning fire aboard! Sail power is clean, silent — steam is dirty and noisy! The engine takes up precious space that can be used for additional cargo. The idea is asinine!" A laughing onlooker commented as the *Savannah* passed on its maiden voyage to England, "It's remarkable that the damn fool contraption works at all!"

The engine aboard the *Savannah* only gave a speed of five-knots at full steam and it was believed its ability to power the ship would only be of use in harbors. Seafarers thought the engine and side paddles had no practical use once the vessel hit the open water. However, Moses Rogers, a longtime captain who saw the potential of steam power for ships, commanded her maiden voyage. He was convinced, that given time, all ships one day would have an auxiliary steam-powered engine, this way sailing ships would never again find themselves caught in the windless grip of the doldrums.

It took the *Savannah* almost 28 days to cross the Atlantic. The engine was used strictly as an auxiliary device (used only 80 hours of operation) rather than relying on its ability to provide primary power to the vessel. The purpose of the voyage was to sell the English on the idea of steam-propelled ships.

While en route, the British bark, *Kite*, observed the *Savannah* passing her starboard side. The lookout aboard the *Kite* noticed smoke coming off her deck and alerted his captain that another

The *Great Eastern*

vessel was seen on the horizon. He yelled, "Look smoke," for surely the passing ship was on fire! The *Kite*'s captain too observed the dense smoke exiting her funnel. He ordered his helmsmen to turn toward the burning ship and offer assistance. "There is nothing worse than a fire at sea. Let's see what we can do to help," he said.

Much to their surprise, it was impossible for the *Kite* to catch the *Savannah*; she was not only under sail, she was also under steam and making a steady 10 knots. The *Kit*'s four to five knots was no match! The *Savannah* disappeared before the eyes of the *Kit*'s crew. The captain later made mention of the sighting before a board of inquiry. He was dumbfounded to hear of such a foolish device as a steam-powered engine although the ship did show great speed.

Once in port at Liverpool, English shipbuilders viewed the vessel as an interesting concept, but thought too it was rather impractical for long sea voyages. Sails were proven and guaranteed not to break down. They had similar comments as did their skeptical Americans counterparts.

The disappointed Captain Rogers returned to America where the engine was later removed and the ship converted back to full sail and used as a coastal freighter running between New England and southern coastal ports.

Several years had passed after the *Savannah*'s initial crossing when English marine visionaries decided to expand on the idea of steam power in their own vessel.

In 1838, English engineers realized, with the proper engine it would be possible to steam across the Atlantic with record setting speed. It was also possible to run directly into the wind eliminating the need to tack. Hence — a shorter route!

The British ship *Sirius* was the first to voyage the Atlantic entirely under steam power in 1838, although she ran short of fuel en route, and was reduced to burning her spars and furniture to continue her journey.

Then, that same year, a larger steam engine was designed and built. After testing the powerful giant it was incorporated into the hull of the new 1,300-ton *Great Western* (236 feet in length) turning two side-wheels.

Cynical observers still commented, "The idea of such a device is foolish and preposterous. A ship of this sort was no more than a waste of time and money."

But despite opposition, the building continued. Finally launched from her ways, she completed successful sea trials. The builders were enthusiastic about the overall performance.

The *Great Western* left Bristol, England, for her port in New York, making an amazing Atlantic crossing in only 15 days. The captain later claimed of running into headwinds during the entire voyage. The ship was more than twice as fast as her sail competitors. He was pleased too that designers placed the ship's coalbunkers holding over 800 tons of coal in the lowest portion of the ship. The heavy ballast provided stability during heavy seas. Coal tenders fed her four large boilers around the clock consuming only 30 tons per day.

Upon reaching the 15th day, witnesses standing at lower Manhattan were awe struck at the sight of the steaming behemoth entering New York harbor.

Once in port, the captain gave a short speech claiming victory to his crossing. Standing before a cheering crowd he said, "The ship averaged a steady 8 knots, the ship's engine never missed a beat — this is the beginning of a new era in ocean crossings."

The captain was right! The *Great Western* continued servicing both countries for the following decade. In total, she made over 60 Atlantic crossings and proved the value of steam power over sail.

She became the first steamer to operate in regular transatlantic service and it was her performance that inspired Sir Samuel Cunard to inaugurate his shipping line in 1840. With the newly devised wooden paddle steamers they made the twice-monthly Liverpool-Boston voyage in an average of 15 days. Cunard's acute business insight was soon followed by other shipping lines. First, the Royal Mail Steam Packet Company which sailed from Great Britain to the West Indies, then, the Peninsular and Oriental Steam Navigation

The *Great Eastern*

Company, which offered service between Britain, India and eventually China.

Shortly thereafter, another engineer who too was fascinated with the success of the oceangoing steamship was Isambard Kingdom Brunel, another Englishman with a vision! His engineering background consisted of mainly applying steam-powered energy to trains running on rails. At the time, railroading was of limited value to Great Britain and government officials cast off his ideas of a national railroad connecting cities of Great Britain.

Thinking his design talents were wasted, Brunel turned his attention toward the sea. He was sure he could improve upon the existing marine designs. Using the now-successful *Great Western* as a prototype, he conceived the idea of another radical design, an all iron hull to replace wood and using a new drive concept, a screw propeller in place of the conventional side-wheels. The hull would be stronger than wood and have no fear of rot or wood decay.

Seeing his ideas, mariners stroked their beards and shook their heads in disbelief. "It is a fool's design, besides iron can't float! Surely, a ship of iron will sink the moment it meets the water!"

Brunel was not only a good engineer, he was an outstanding salesman. He met with English businessmen and explained his design. Using charts, graphs and handmade models he sold the unique concept and generated the financial support necessary to fund such a program. The man proved his theory to be correct.

In 1845 the 1,300-ton, 320-foot *Great Britain* was launched, boasting of six-unsinkable watertight compartments within her all iron hull. The ship's powerful steam engines developed an amazing 1,200 horsepower and turned a huge 15-foot diameter screw propeller, another radical design improvement. Brunel estimated, the ship would reach a possible eight or nine knots in all forms of weather. On her first trial run, the ship steamed past time clocks at over 11 knots!

Competing engineers still cynical of his ideas thought the design and speed of the *Great Britain* was over exaggerated and no doubt false. The posted figures were strictly a ploy and used to impress investors. "No iron ship could withstand the pounding of continuous sea duty without sinking and the speed was an obvious error in calculation."

Again, Isambard Brunel proved them wrong. The *Great Britain* remained in service for over 60 years until she eventually ran aground just off the Falkland Islands.

The inventive engineer envisioned still another great ship, only this time bigger, faster and more luxurious —a vessel displacing an enormous 27,400 tons, 692 feet long to be launched in 1858. Named the *Great Eastern*, Brunel came up with another successful huge screw propeller measuring a whopping 24 feet in diameter to drive the all-iron hull. Its blades stood four times the height of a man! He also managed to bolster the ship's steam engines to develop 1,600 horsepower. The *Great Eastern* retained six masts and an elaborate suit of sails, but she also carried a set of paddle wheels run by one steam-engine and the propeller run by another Brunel innovation, a separate engine to power her steering.

Unthinkable for its day, she later carried a crew of 400 and 4,000 passengers, and remained the largest ship in the world until the end of the 19th century. However, the entire concept during the late 1850s was unheard of and considered way ahead of its day! Nonetheless, Brunel, having shown the former success of the *Great Britain*, found little trouble developing new financial backers to pay the enormous building project. The ship's keel finally was laid. But unforeseen disasters fell just ahead!

Hundreds of spectators joyfully stood by waiting for the new ship to be launched in 1858. Well-wishers patted Brunel on the back, congratulating him on developing such a vessel.

Due to her extensive weight, fellow marine engineers who worked alongside Brunel thought best for the hull to be laid running parallel to the Thames River rather than sending her into the water stern first. They feared her enormous size and weight would be too great and the ship would dig directly into the riverbed.

The signal was given to launch. As her massive hull began moving seaward, the hydraulic supports suddenly gave way. The extreme weight of the ship caused the hull to slip off the vertical stocks. Then in a quick succession, the restraining lines snapped, causing the hull to freeze nearly where she was built. Workers fought frantically to move her off, but their efforts proved unsuccessful. The *Great Eastern* remained stuck on the ways as the disappointed audience stood watching.

For days thereafter, shipwrights worked frantically to move the giant ship. Nothing seemed to work. It wasn't until an unusually high tide appearing several months later that workers found a way to permit the big ship to move and finally take to the water.

The bad launching and expensive overrun costs eventually took two more years to complete the inner construction. Facing financial ruination, the original backers lost interest and folded the company thus, ending their high losses. The big vessel now sat dormant tied to a wharf.

Convinced he was still on the right track of building a ship of such magnitude, Brunel went back to work in search of new financial support. Sarcasm prevailed; many thought the man was a better salesman than an engineer! Whatever his secret, he managed to find new money to support completion of the ship.

Problems and frustration continued, and the men working aboard building the ship's innards began talking of the devil working at their side! Now not only were there financial excesses, there were dire labor problems to contend with. The ship seemed plagued with construction mishaps — without provocation, several men were killed in unexplainable accidents. Workers also claimed they could hear strange tapping sounds stemming from the parallel walls of the hull. Apparently, a riveter working on the inside of the hull mysteriously disappeared and could not be accounted for! Fellow workers thought the man was entombed within her hull. Now, a ghost also haunted the ship! Then the worst news came! The designer, Isambard Brunel, died from an undisclosed disease.

Despite all of the mysterious acts that shrouded the ship, the completed *Great Eastern* made her maiden voyage in 1860. She provided transportation across the Atlantic for several years. She operated unscathed and was considered the luxury liner of her day, but failed to make enough profit to cover her high operating costs. Owners were devastated by their continuing losses.

Then, in 1865, she found a new life— due to her enormous size she was the only vessel capable of laying the first 2,000 miles of the transoceanic telegraph cable between Europe and North America.

Irony called again! Just after placing nearly 1,200 miles of the heavily insulated wire that measured nearly two inches in diameter, it snapped dropping the untied end into the sea. For days that followed, the hard-working crew grappled for any portion of the cable. Wind and high seas dispelled any further hope for retrieval. They returned to port and waited for better sea conditions.

Another year passed, finally success prevailed in July 1866. The cable was again grappled in pounding seas. After several days of frustration by the crew, the end was found, attached and brought ashore in Newfoundland. The ship continued as a cable layer working worldwide until 1873.

By 1889, the owners felt the *Great Eastern* had outlived her usefulness. She accomplished what no other vessel of her day could perform — laying communication cables throughout the world. Salvagers took their torch to her hull and she disappeared without a trace.

It took considerable amounts of money, much in the way of unique design and years of convincing that steam-powered vessels was a safe and practical mode of propulsion. Steam-power finally turned the corner of ships that crossed the vast seas that lay about us.

A Whale Of A Story

Sag Harbor, Long Island, New London, Connecticut, Nantucket, New Bedford, Massachusetts — the names were different, the ships that anchored within their small ports were the same. Each berthed square-rigged sailboats of assorted sizes. Ships that went to sea for years with no guarantee of a safe return!

But why? What was so all-inspiring about getting on a vessel and looking for a huge monster that roamed about the sea?

The 1840-1850s was considered the height of the whaling industry. It flourished in New England with little competition. Ship's masters sought great interest in harpooning the mighty whale — few animals provided so much use and/or value to our growing civilization. And upon return, all who worked the decks of these ships were usually assured great riches.

Able-bodied seamen would see $500 to $1,000 (less onboard expenses) the captain, $2,500 to $5,000 and the ship's owner, an astonishing $25,000 or more! Needless to say, for this pay a voyage normally lasted several years. Huge sums of income considering the day — common laborers often made only a dollar daily wage and survived!

These unique vessels varied in size from 80 to 150 feet in length. Most carried square-rigged double masts of bark design. While at sea, each vessel was equipped to take aboard a large volume of these animals (eight or more depending on the size) then process the remains into several by-products. Masters would only return to homeport when their ship's hold was full.

Whalers traveled worldwide depending on the season of the

year. Captains were well schooled in the lifestyles of whales and knew their migrating habits. In general, ships followed the herd wherever it took them.

Local communities aforementioned depended heavily on these ships for their basic economy. Townships provided the services necessary to support life at sea. Carpenters, sail makers, toolmakers, blacksmiths, shipwrights and even local farmers were considered as crafted individuals who kept the fleet afloat. The ship's thirst for stores — wood, rope, barrels, tools, hardware, canvas to name just a few — added to the financial support of the neighboring towns.

Finding qualified seamen proved to be the most difficult chore within the industry. Young men were wooed by the potentially large income. It wasn't long after leaving port that working conditions were discovered as appalling! Most seamen were only boys in their late teens, lured aboard by unscrupulous masters.

Whalers had a terrible reputation — once at sea, surely long hours of hard work were expected, harsh treatment by their masters and voyages lasting for several years before returning to homeport. Lay seamen considered the ship no more than being thrown into a seagoing jail!

A whaler's crew consisted of no fewer than 30 men to man the vessel. In addition to her master, there was a group of first-mates, normally one assigned for each of her whaleboats, also a harpooner for each boat, a cook who acted as steward and at times ship's doctor. The balance of men were deckhands and ship-handlers.

Life at sea was considered miserable! Bad food, cold damp berths, or siphoning heat in the warmer waters. Bedbugs, roaches and lice were a constant menace. Rats roamed about in bilges and cabins, generating nauseating odors within the Stygian quarters.

After leaving homeport, the whaler looked for nearby trade winds that took to the ship to the deep-water nesting grounds of their prey. There, the hunt began!

Lookouts were posted for the first sign of a whale with its flukes rising out of the water or the sight of water rising from the mammal's blowhole. Usually, the observer perched at the highest

A Whale Of A Story

point on the mast had the best opportunity of a first sighting!

When the cry, "Thar she blows off the port bow," was heard when a whale was sighted, adrenaline began running high in all no matter who they were aboard ship. Seamen immediately ran to their assigned station. The command was given; sails were hastily lowered, small whaleboats deployed from their block and fall.

The tiny boats were manned by a crew of oarsmen, the harpooner and a first mate who was in charge of the hunt. With a sharp eye and direction usually called from the mother ship, crewmen rowed their tiny boat dangerously close to the creature until the harpooner could throw his dagger-like harpoon into the back of the beast.

With the animal obviously now in severe pain, it would take several dives to somehow try to remove the dart like object from its back. Extreme care had to be taken by the first-mate as he guided the helm of his whaleboat to avoid being hit by the thrust of the whale's enormous flukes. The animal would dive for deep water then surface rolling and tossing its painful body. One swipe of its giant fluke could easily tear the boat into pieces killing the entire crew. Very often, accompanying boats would join the hunt, thrusting their harpoons to speed the killing process!

If the harpooner missed his target, very often a whale would live for several hours, dragging the group of hunters behind. The whale often groaned in pain or sent a screaming sound, placing a chill down the spine of young sailors. While in tow, the men could do no more than hold on for dear life, praying the beast would tire and die quickly! This was also referred to as the "whalers' sleigh ride!"

With the dead animal in tow, oarsmen returned to the mother ship. Weighing up to several tons, it took all hands rowing with maximum effort to bring the dead carcass aboard. Despite being exhausted from the killing chore, this was no excuse for the crew to slack off from the workload. Everyone was expected to provide his share of work until the animal was completely consumed and stowed beneath decks.

A large wood platform was lowered and secured to the hull.

Large blocks and tackle were tied to the whale, providing direction as it floated on. Once the body was secured, the cutting process began. Seamen using sharp-spade like knives sliced the blubber. Large pieces were hauled on deck via a windlass; its oil was boiled off then place in kegs. It made little difference how long it took; the master expected results, often making immediate made plans to continue his search for the next prey.

Often as the cutting continued, whale blood and pieces of waste fell into the water, causing schools of sharks to form about the ship. Crewmen wielding their long cutting knives stood cautiously on the slippery planks to avoid falling into the shark-infested water.

The animal's extract provided a generous supply of lamp oil and was considered a high-quality lubricant for fine mechanical devices such as clocks and machinery. The flexible properties found in whalebone were used for men's shirt collar stays, ladies' corsets, skirt hoops and whips. Ambergris, a gray waxy substance secreted from the intestines of the sperm whale, became the base of expensive perfume. And spermaceti, a derivative of the whale's head, went into the making of candles and ointments. Few portions of the mammal were wasted or discarded to the sea.

Several cases of mutiny were reported due to long voyages, disease, bad food and severe punishment by a ruthless master. Whalers sailed to all points of the compass — very often finding themselves thousands of miles from homeport. On occasion, the captain would sell all of the by-products while replenishing stores, keeping majority of the money for himself. Under such circumstances, the crew mutinied! Fearing reprisal by authorities, very often the ship would sail on, her master killed and officers vanishing mysteriously from the record books.

During the 60 or so year heyday of whaling, thousands of men failed to return to ports along the New England coast. Granted, records also showed losses at sea due to bad weather, hurricanes and typhoons plus unaccounted sea tragedies of fire and sinking. But the majority of disappearances were due to mutiny, whereby men found refuge on South Sea islands. Friendly islanders welcomed the white-face strangers providing them with shelter and willing women to service their personal needs.

As a rule, after several years at sea and their holds filled to maximum capacity, the whalers returned to homeport. Onlookers and family members waved, cheering the men home as the ship made preparations to dock. Tears formed in the eyes of men who saw their child for the first time — often their offspring was two or more years old!

The ship was unloaded, its treasure sold to the highest bidder and preparations were made to return again to sea. Ship's crews rarely saw fellow crewmen again! "One voyage aboard this old tub is enough for any man," said a weary seaman getting off. Rule of thumb, only senior officers and ship master's returned to the hard lifestyle.

The demise of large-scale Yankee whaling was brought to its knees in 1859, when "black-gold" (crude oil) was found in Titusville, Pennsylvania. Oilmen suddenly found they were able to pump crude easily and showed itself as an endless supply! There was no telling how large the pools were in the ground. Titusville oil was not only equally as good; it was considered a better, cheaper fuel than its whale counterpart.

It wasn't long thereafter before the whale oil business crumbled — many ship owners found no consumable market! Unsold, untapped barrels were found in warehouses months after processing.

Entering the American Civil War was no doubt the final blow for the whalers. During the height of the war in 1863-1864, Confederate raiders took their toll on Yankee whaling ships. More than 70 vessels calling New England ports their home were seized and burned by the South.

The North sought revenge to their horrendous act and towed several of its older ships to southern ports. There, ships holds were torn open and flooded letting them sink with the hope of blockading harbor entrances. Their attempt failed and the majority sank in deep-water as planned doing little harm.

The whaling industry ebbed slowly to a close. Several ships

calling New England ports home continued making extended voyages, but the need for whale products declined as the 20th century turned the corner.

Only a handful of countries now support the whaling industry, most of which formed a union limiting the amount of whale kills annually. The American government no longer permits any form of whaling: by contrast, it prefers to cultivate the herds that remain.

Today, the only direct remains of the industry are displayed at the Mystic Seaport Museum in Mystic, Connecticut. There, the 105 foot *Charles W. Morgan*, a whaler built in 1841, is beautifully restored. She is the only reminder of an era that has long been forgotten.

Christopher Columbus, Salesman!

Christopher Columbus should no doubt be labeled the best salesman of his day. He had a vision of finding a new path to Asia that was unproven and had to be sold! Although obviously there were no photos, several artists painted the explorer as a man who stood about 5'10" tall with red hair. His face was drawn yet they painted him having sincere looking eyes. Born in Genoa, Italy, in 1451, he had little education and learned to read and write on his own. Later he married Dona Felipa, who died just after giving birth to their son Diego in 1480. History shows, his father was a weaver and thought to have had two brothers.

As did most young men of Genoese heritage, he voyaged in the Mediterranean learning seamanship and navigation skills. In 1476, his ship was wrecked off the rocky coast of Portugal. He managed to find his way ashore and went on to Lisbon. Boarding another ship, he claimed of sailing to northern ports in Ireland and England and later to have gone as far as Iceland!

The man was God fearing, using Scriptures from the Bible for guidance. He spent hours reading the testaments, interpreting their meaning. One passage in particular from the Second Book of Esdras read: *Thou didst command the waters to be gathered together in the seventh part of the earth; six parts thou didst dry up and keep.*

According to historians, these few words were the inspiration that led his thinking there was a direct sea route to the Orient. If the sea actually rested on a globe and it represented only one seventh of the world, Japan and all of her riches were only a short sail to the West.

Most friends considered him bright and self-taught especially when it came to science. Although few could refute his ideas most were still superstitious of falling off the end of the earth. Making passage west was impossible!

Columbus's fascination grew after finding Scriptures by the ninth-century Moslem astronomer, al-Farghani. Using the stars as a point of reference, it was he who calculated a degree of longitude was the equivalent to 66 nautical miles. With this figure in mind, using the Canary Islands as a starting point, Columbus calculated the Orient only lay 2,400 miles to the west. He also calculated, at a sailing speed of 100 miles per day, the voyage would only take 24 days to complete. What was not realized, a severe navigational error was induced from the misinterpretation of al-Farghani figures by the explorer. Columbus used 45 miles in lieu of 66! This gross error placed Japan on his map. In truth, the location was the West Indies.

In addition, he studied Marco Polo's maps which later proved erroneous. It located Japan 1,500 miles east of China — and Ptolemy's underestimation of the circumference of the Earth and overestimation of the size of the Eurasian landmass. Columbus thought Japan was about 2,400 miles to the west of Portugal — a sailing distance obtainable by existing vessels. His idea was furthered by the suggestions of the Florentine cosmographer, Paolo dal Pozzo Toscanelli.

In 1484, convinced his theory was correct, Columbus placed his facts before a formal panel of scientific experts including John II of Portugal, grandnephew of Henry the Navigator. After listening to these wild ideas, each shook his head in disbelief. Columbus" theory was insane and surly any man making a voyage of this nature was making no more than a death wish for himself and those who sailed with him. The panel refused his ideas and calculations, believing the distance to be beyond the capa-

Christopher Columbus, Salesman!

bilities of existing ships.

Then, in 1485, Columbus took his son Diego and went to Spain, where he spent almost seven years trying to obtain support from Queen Isabella of Castile. Eventually, he was received by the queen gaining friends as well as enemies. In 1492, faced with an apparent final refusal, Columbus prepared to take his proposal to France where he thought he could sell the royal court on his pending ideas.

Although he was rejected by his peers, Columbus" determination persisted. He was still convinced in his calculations that the voyage remained plausible. Taking his proposal to Queen Isabella for the last time, the mariner began reselling her on the idea. Opposition remained; her advisors pleaded that the man was still wrong in his thinking. Columbus proved indeed to be a good salesman! Isabella overrode her advisors and granted the man permission — the Spanish crown would supply three vessels.

His flagship was the 90 foot *Santa Maria*. Two support caravels, the *Pinta* and the *Nina*, were outfitted with supplies in the tiny port of Palos. Columbus was aided in recruiting a crew by two brothers — Martin Alonzo Pinzon, who received command of the *Pinta*, and his younger brother Vicente Yanez Pinzon, who commanded the *Nina*. The ships assigned crewmen consisted mostly of thieves, rapists and murders released from jails by royal amnesty plus a small handful of professional seasoned seamen who were ordered aboard by command of the queen.

The three tiny boats left Palos on August 3, 1492. Seeing the *Nina's* poor seaworthy condition as she followed the *Santa Maria*, Columbus ordered the *Nina* back to port to be rerigged in the Canaries. He was sure her poor sail design would fail in the expected western seas. It was from this point the three officially sailed to the west.

From the start of the voyage superstition ran rampant throughout the crews aboard the three tiny vessels. A majority of the hands believed this was a journey for the doomed. Bad omens showed daily signs of tragedy; sailing on Friday, spilled salt and crossed knives were found on tables, even a sneeze from fellow seamen who just boarded the ship! All signs of future disaster!

Days passed, then weeks as the ships beat into a strong winds. No sign of land in sight! Columbus took daily computations giving reduced figures to his men of the distances covered. He feared providing the crew with actual mileage covered would indicate the return again to Spain would be too great! Then, to worsen their situation, the prevailing wind stopped — the three found themselves becalmed for hours on end. Still another bad omen.

Food ran low; water turned, forming a rancid green slime in water kegs. On the day of October 10th, there was murmur among the crew of mutiny! Surely they would perish if they did not go back. Their captain was a madman!

Columbus pleaded with the men, promising only a few more days before reaching their goal. Now in a desperate situation, the first signs of life appeared, providing credence to the master's words. Tiny pieces of driftwood bearing fresh leaves were gaffed from the sea. It was the first sign of hope that land was within reach.

On the morning of October 12, 1492, the observer standing watch on the *Pinta* saw what he thought was a gray cloud sitting on the horizon. Moments passed, then convinced the sighting was not a mirage, he gave the call of sighting land. The three small ships finally made landfall in Japan!

Proudly rowing ashore in boats, crewmen were greeted by naked, dark-skinned tribesmen who at first were timid. The inhabitants, Arawaks, were a friendly local population of natives that Columbus identified as Indians. Looking about the inviting white sand, there was no civilization, no buildings or docks to accept their boats. Surely, this island lay offshore from mainland Asia. In reality, it was an island in the Bahamas, which Columbus named San Salvador.

Days later, the expedition sailed farther west to another island, Cuba, where a delegation of his crew was sent to seek the court of the Mongol emperor of China and the gold riches the land had to

Christopher Columbus, Salesman!

offer. Again, surprise! The men were greeted by more overjoyed Indians who thought Columbus and his crew were descendants from heaven. They were welcomed with open arms.

After replenishing their holds with new fruits and strange foods, they sailed east in December to Hispaniola. It was Christmas when misfortune struck. The *Santa Maria* was shipwrecked on unforeseen rocks near Cap-Haitian. Columbus managed to get his men ashore unscathed. The admiral decided it was time to return to Spain. Thirty-nine men were selected to remain on the island at the settlement named Navidad.

Reboarding the *Nina*, the crew set sail for Spain leaving the small contingent of men behind with the promise of more ships and men to follow. Martin Pinzon, who had explored neighboring islands on his own with the *Pinta,* rejoined the *Nina* and followed. During the return voyage, the ships were separated by a fierce storm that nearly sunk the *Pinta.* The *Nina* continued on.

In March of 1493, the *Nina* made port in Lisbon. The admiral was hailed as a hero and greeted joyfully by John II.

Crossing Spain to Barcelona, Columbus was welcomed in triumph by Isabella and her husband, Ferdinand II. Those adversaries who claimed he was insane now hailed him as a fellow monarch. At their first meeting, Columbus claimed to have reached a group of islands just off the coast of Asia. He explained on future voyages, they would find the mainland of Asia where gold and pearls were to be found. His talk was so convincing, no one reputed his idea of a fast return.

The Second Voyage

Generously funded by Ferdinand and Isabella, the admiral grouped 17 new ships and almost 1,500 men for a return voyage to the New World. The flotilla sailed from Cadiz on September 25, 1493, making landfall in November 1493, near the Lesser Antilles.

Their expedition continued and additional islands were discovered, named and charted. The explorers moved on to the original site of Navidad where they had left the 39 men from their first voyage.

Discouragingly, they found the tiny village of Spanish sailors destroyed and the men killed by local Indians. It was later uncovered the seamen had pillaged the island and raped all the women.

The admiral organized a new colony, named Isabella, about 75 miles to the east of the Navidad's site. Christopher Columbus tried to govern the colony until he returned to Spain in 1496. To his misfortune, he was found to be a failure in government administration. Ambitious underlings assigned as aides underminded his authority and complained about his leadership ability in Spain.

He left his brother Bartolome in charge with instructions to move the settlement to the south coast of Hispaniola. In 1496, the settlement was named Santo Domingo, later becoming the first permanent European settlement in the New World.

Columbus returned to Cadiz in June 1496. While in court, adversaries again took a dim view of his not finding the rich Asian mainland. All attempts to obtain gold from the Indians on Hispaniola had been unsuccessful.

The Third Voyage

The old explorer was finally authorized to make his third voyage after the Portuguese had sent Captain Vasco da Gama off to India in 1497. Columbus departed with six ships in May 1498. Fighting strong headwinds, the ships made landfall on the island of Trinidad mid-July 1498. After a week ashore, his tiny fleet continued sailing west where they finally reach the mainland of South America.

Having found traces of gold near the coast, their expedition crossed the Caribbean to the island of Santo Domingo. Much to his surprise, many of the new colonists there were in revolt!

Francisco de Bobadilla, the island's new commissioner, was granted full authorative reign by Spanish royal authority. He removed Christopher Columbus and his brother, Bartolome from government service and sent them back to Spain bound in chains. Devastated by the commissioner's action, the brothers stood before Ferdinand and Isabella still in shackles. The king and queen were appalled and ordered both set free.

Fourth Voyage

Still determined to find passage to Asia, Columbus soon set forth another expedition, which sailed in May 1502. Arriving in Santo Domingo, he was denied permission to land.

Taking heed, his ships sailed along the coast past Panama, finally returning again for Santo Domingo. He made note they were within 10 days of sailing to the coast of Asia! Again faced with denial, he sailed on. However, time and rough seas took their toll on the small ships. All his vessels were in poor condition. With worn, tattered sails, rotting hulls and shredding lines. Crews feared sinking at sea if something was not done. Finally, arriving in Jamaica, the decision was made to abandon the idea of going on. The admiral and his men found themselves marooned for almost a year while they awaited rescue — help did come. He finally reached Spain in November 1504.

Upon arrival, he found he no longer had the support of King Ferdinand and Queen Isabella. Much to his dismay, the Spanish crown had, from 1495 onward, violated its original agreement with the old man by authorizing others to sail to the Indies.

Christopher Columbus died in Valladolid on May 20, 1506, at the old age of 45 years! The admiral's genius lies in the fact that it was he who found the land now known as the West Indies. Despite having made major errors in navigational computations and location, he was able not only to find his way back to Spain, but also to return to the Indies time and time again.

Columbus was indeed a great salesman, a visionary who sold his country on a radical idea. His original 33 day triumph will be remembered forever.

Sea Search

The *Eastland*, Perished Where Berthed

While berthed at a Chicago River pier just off Lake Michigan, a company greeter stood at the rail welcoming passengers aboard the *Eastland* day line steamship.

"Good morning madam, lovely day for a sail! Welcome aboard folks, please watch your step!" were the words of the smiling employee as he assisted several of the ladies wearing long dresses and carrying parasols over their shoulders. Well-dressed passengers held children by the hand returning a smile or nod as they looked for friends and relatives already aboard. Walking the promenade decks, no one realized they were boarding a ship that would in minutes to come, claim 845 lives and become the worst marine disaster ever recorded on the Great Lakes.

Known as the Queen of the Lakes, the *Eastland* was considered a favorite excursion steamship as she carried passengers on day cruises along Lake Michigan's Chicago coast.

It was a typical warm Sunday morning, July 24, 1915, when the *Eastland,* measuring 270 feet, waited dockside on LaSalle Street. Some 2,510 passengers stood in line to board; many of the smiling faces were from local businesses that co-sponsored the day's outing. Men and women joked, laughed and stood patiently as the ticket master checked each for boarding. In typical fashion, children ran about playing hid and seek waiting to be called by their parents.

Despite the gleaming paint and handsome silhouette of the *Eastland*, there were hints by competitors that she was an unsafe cruise ship! Many of her adversaries claimed the vessel was not seaworthy due to poor marine design. She was considered unstable and top-heavy. Without the proper ballast over her keel the ship would surly topple while at sea.

The ship's owners were outraged with the unfounded accusation. In fact, they were so appalled that the owners ran an advertisement in local tabloids offering a $5,000 reward to any marine design/engineer who could substantiate the claim of the ship not being seaworthy. In their own defense, the ad also boasted, the *Eastland* carried over 200,000 passengers without any sign of trouble or marine incident!

Although the controversial publicity remained, few passengers paid attention to the demeaning claims. After all, it was just the cry of competition!

The signal was given that all were aboard and that the gangplank be hauled on deck. Dock lines fore and aft were dropped. The ship's whistle blew and a small tugboat began to tighten its towline as it began to pull the *Eastland* from its berth.

According to reports, a majority of the passengers stood on the port side of the vessel. Men waved their fashionable straw hats and ladies shook their scarves to onlookers. Without warning, the vessel's hull started to list slightly to port. Few paid attention to the unusual attitude of the ship. Only seconds passed when wood deck chairs and untied furniture began sliding toward the opposite railing. Someone on deck yelled, "Look out! We're going over! The ship is going over!"

Screaming passengers looked frantically in every direction to find anything stable to hold on to while the giant hull began to roll on her port side. Tables, chairs and loose debris slid along the slanted deck slamming into passengers who clung to grab rails. Those unfortunates who stood mid-ship and on the starboard side slid along the tilting deck. Losing their grip, horrified passengers fell into others and then into the muddy water of the Chicago River. The deck angle became so acute, it was now impossible to maintain a hold on the hands of children and spouses.

One onlooker reported, "People began falling into the water in droves yelling and screaming. They were helpless; there was nothing to hold on to! Then, others fell on top of them pushing their bodies under the murky water. What a God-awful sight!"

The rolling hull finally stopped resting on her side in 25 feet of water. Another 20 feet or so of the hull remained above the water's surface. Hundreds of horrified passengers clung to the opposite rail trying in desperation to grab a footing onto the exposed hull and remaining angled deck. Within less than a minute, hundreds of woebegone bodies (soon to become over a 1,000) found themselves groping for anything that would float or help save their lives. Most individuals were not capable of swimming. Those who could were pulled under by others who floundered or groped for life. Women wearing multi-layered undergarments found their clothing acted as if a sponge were wrapped about their bodies. "It appeared as they were tied to anchors," remarked still another aghast onlooker.

Others, who had arrived early went below decks and sat happily in the confinement of their cabins. They found themselves trapped by the cascading rush of water flooding the bowels of the vessel. Cabin doors flung open as a wall of water filled their rooms quickly drowning all who remained. The capsizing of the *Eastland* happened so quickly, few passengers had time to respond to any form of a safe egress.

The word spread quickly along the pier with people shouting, "The ship just turned over, get help someone, get help!" Bystanders looked horrified not knowing what to do! The ship now rested only a few feet from the pier. The poor souls floundering before them were practically within arm's reach, yet there was little anyone could do to help. They stood helpless watching friends and relatives drown before their eyes.

Water poured into the engine room snuffing out the boilers. In so doing, huge clouds of steam formed causing harrowing belching sounds to emanate from within the hull. The rollover happened so quickly it was impossible to get the men and under-deck passengers out from the grasp of the ship.

Seeing the tragedy unfold before their eyes, neighboring tugboat captains ordered their boats alongside to aid in the rescue. Deck hands used peaveys and/or lifesaving rings to grab the few who survived the accident. Captains found it almost impossible to maneuver their boats within the close confined space. They too stood in horror as bodies disappeared before them.

One of the survivors, Martin Birch, gave the following accounting of his experience: "I just came aboard with my wife and young son. We were directed to a small compartment on the starboard side of the ship. My wife suggested that I change from my work clothes into something more appropriate for the cruise. As I began to undress, the cabin seemed to be moving somehow then furniture shifted toward the door. Water in the wash basin began to pour onto the table. My wife looked at me. 'Martin, something is wrong here! The room seems to be tipping!' She no sooner finished saying the words when we heard screaming from the halls and other rooms. A man in the hall yelled, 'Good Lord, the ship is about to capsize. Get out before we're all killed!'

"The next thing I knew there was a loud crashing sound; I think it may have been something in the boiler room. Then, all of the room's contents fell against the cabin door. We looked at each other trying to hold on to anything to keep from falling over. I knew we were in serious trouble. There was more screaming and the continued cry, 'Get out! Get out!' Although we were standing at right angles to the walls, I did manage to pry the cabin door open. People were walking on the companionway walls! Steam or possibly smoke began to fill the halls. It was a strange sight.

"With the help of another man, (I'd like to know his name!) he helped the three of us to climb the stairs and work our way to the upper deck. Walking, I remember being very disorientated and somewhat dizzy. Nothing made sense. My wife held on to my coattail, my son held her skirt. There was little doubt the ship had rolled over. Mayhem ran throughout as people did not know what to do. I realized we were lucky; our portion of the ship was still above water. Several people staggered out of partially submerged doors. I was sure many were still trapped on the other side of the ship. Those who could no longer maintain their grip let go and slid across the deck into the water. While the three of us were clinging to an exterior cabin wall, my wife looked at me then said, 'Martin, you forgot your trousers!'

"It was an ordeal I never want to experience again. We waited for almost an hour before being helped to safety."

When the tragic few minutes was over, a total of 845 perished in the accident. Hard-hat divers retrieved bodies days after capsizing. Upon entering the closed compartments remaining underwater, divers found it most disheartening to discover entire families drowned — many still clinging arm in arm to loved ones.

Unbelievable to most, the poor individuals who perished were only within several feet from the dock. There was no fire, no ramming while at sea and the stormy weather normally associated with sea tragedies was nonexistent.

Months later, the closing inquest finally determined the ship capsized due to the poor arrangement of ballast within her keel. There was no weight compensation made or allotted for having so many individuals on one side of the ship. Engineers later calculated that nearly 1,000 persons no doubt stood along the port railing of the *Eastland,* causing the capsizing. At 175 pounds per person, (average weight), that equaled 88 tons of human body weight not compensated for with the lower ballast limits. The event happened so fast there was no time to redistribute the unforeseen imbalance.

Later, competitors smugly sat nodding their heads, "See! I told ya so."

Her name was the Arabia, *a steamboat found traveling along the Ohio, Missouri and Mississippi rivers.*

The *Arabia*, She's A Steamboat A' Comin'!

The sound of her steam engines and swooshing of her paddles were an undeniable sound a riverboat was moving along the river. Then, as though it offered a sign of welcome, the captain would blow the throaty steam-driven whistle letting everyone along the banks know they were en route to a nearby quay.

During the early days of our migrating west, young children stood waiting patiently along riverbanks experienced this excitement. There was joy seeing two towering stacks spewing heavy black smoke, paddles churning effortlessly and the growl of the steam engine working within its innards. Then, their cry informed all, "Steamboats a' comin', steamboats a' comin'!"

To the newly formed villages, the riverboat was the lifeline of progress bringing people, tools, food, machinery and merchandise of every description to a land that was growing.

Her name was the *Arabia*, a steamboat found traveling along the Ohio, Missouri and Mississippi rivers. The John Snyder Pringle Boat Building Company built her on the bank of the Monongahela River at Brownsville, Pennsylvania.

Her keel was laid in 1853; she was 171 feet long and displayed a beam of 29 feet plus her side-mounted paddle wheels. According to her builders, she would carry some 222 tons of freight within its hold and on the planked decks. Each of her 28-foot-diameter paddle wheels was powered by two individually operated steam engines. The boilers were of iron-jacketed three-tube construction. The powerful engines consumed up to 30 cords of wood per day depending on speed and load.

Pringle and his boatwrights were so skilled in their trade they boasted the company was capable of producing a finished vessel in less than six weeks' time. They became so successful in selling riverboats; the company ultimately took control of a nearby sawmill and an iron foundry producing state-of-the-art steam-engines for river craft.

After her initial launching, the *Arabia* worked her way to the Ohio River, the Mississippi and finally to the muddy Missouri. During the ensuing years the ship was bought and sold several times — ending with ownership by Captain John Shaw of St. Charles, Missouri, and two partners, George Boyd and William Terrill, both from St. Louis.

It was Shaw who found an excellent customer to hire their ship — the U.S. government. Soldiers and munitions were needed to support remote frontier outposts situated along the upper Missouri River. The much-needed supplies were ultimately delivered to the base of the Yellowstone. The trade was lucrative and its steady flow provided a stable income for the company.

The ship's first encounter with disaster appeared after leaving Ft. Leavenworth, Kansas, in March 1856. The ship made her usual river run carrying men, munitions and supplies. During the trip, the *Arabia* found herself dangerously close to the river's edge due to a heavy ground fog. Although the captain ordered reduced speed, the helmsman misjudged his distance — the hull encountered jagged rocks resting in a turn of the river. Not realizing the close proximity to the shoreline, the ship tore a large hole in her hull and without warning ripped off the tiller. Knowing his inability to steer, Captain Terrill was faced with a desperate situation and ordered the ship be run aground. His decision was wise, for when the vessel finally stopped along the riverbank, more than two feet of water had already filled her hull destroying much of the perishable cargo below.

Now, miles from civilization and or any form of aid, the captain and his crew were forced to improvise and make a rudder using wood from a nearby tree. Only hand tools were available, making the work slow and arduous. Makeshift planks were cut then set within in her hull to seal the gaping hole. Finally, hand operated bilge pumps removed enough water to safely refloat the ship. The repair took nearly two weeks to complete before they reached port in St. Louis.

Several more trips were made that year feeding welcomed supplies to frontier towns along the Missouri. Then, late in August, tragedy struck again at the Quindaro Bend, near Parkville only a few miles north of Kansas City.

Their final destination was Council Bluffs, Iowa, with interim stops servicing tiny settlements along the river's edge. The ship's hold and decks were filled to near capacity with freight of every description along with livestock and paying passengers.

As the *Arabia's* side-wheels turned effortlessly in the muddy water, Captain Terrill stood watch in the wheelhouse checking the ship's position along the river. It was just after sunset, and the stewards were busy serving dinner. Without warning, the *Arabia* experienced a loud crunching sound followed by a vibration stemming from the bow.

Evidently, a large tree floating partially submerged acted as though it were an underwater spear, piercing the ship's wood hull while she moved at near full speed. Captain Terrill ordered immediate reversal of the engines, but it was too late. The tree trunk and its water-soaked branches had tore a gaping hole in the bow permitting a cascade of river water to enter the forward compartment. Within minutes, the bow began to slip into the river.

With the forward deck now submerged, water quickly flooded the engine room, snuffing its boilers. It was too late! Not only was the ship unmanageable, it was sinking too! The ship's master knew his vessel was lost.

Apparently, the keel found a temporary seat on the muddy riverbed, causing the ship to list heavily to port. Furniture of every description, wood crates and food trays slid along the floor. Passengers had all they could do to avoid being thrown against cabin walls or into the river itself.

Realizing their desperate situation, several quick thinking men ran to launch the only available lifeboat from its davits. The small boat was a true lifesaving device making countless trips back and forth removing women and children from the tilting decks.

In the wheelhouse, the belching horn was continually sounded, informing those on land of their desperate plight. Seeing the perilous situation of the ship, nearby residents assembled teams of horses pulling wagons to assist in the ship's evacuation. Stricken passengers were taken unscathed to a nearby hotel in Parkville where they could rest for the night. Crew members remained behind to help unload all salvageable cargo.

Now, with passengers safely off the vessel, local townsmen began removing personal effects from the drowning ship. Steamer trunks, assorted luggage, parcels and created freight containing rifles, clothing and food were handed from man to man then loaded into waiting wagons. Unbeknownst to most, the wagons were driven off into the night then pillaged by neighboring thieves who rummaged through every piece removed. To their misfortune, the gold and silver coinage supposedly placed within the ship's safe could not be removed in time. It slipped beneath the muddy water and was impossible to retrieve.

By the following morning, shoreline bystanders were amazed to see the sinking remains of the *Arabia*. Only the tops of the pilothouse and the two smokestacks were visible above the murky water. The river somehow seemed to swallow the balance of ship and all its riches.

The headlines in the September 6, 1856, *Kansas City Enterprise* read: *The Steamer* Arabia, *bound for Council Bluffs, sturck a snag about a mile below Parkville last night and sunk to the boiler deck. — Boat and cargo total loss.*

Later revealed, the real treasure aboard the ship was not its coinage and freight, but its 20,000 gallons of Kentucky bourbon whiskey resting in nearly 400 wood casks. For years that followed, salvagers considered this prize as the most valuable of all.

Time passed and the vision of the ship faded in the minds of most. For those who lived within the near proximity, the story of the tragic sinking became no more than a barroom story. As with any key event, storytellers added their interpretation of what happened during the sinking — usually adding fictitious anecdotes to enhance the flavor of the yarn!

Nature too took is course and began to make a natural change in the direction of the Missouri River. Its flowing path changed due to time and the natural erosion process. The final resting place of the *Arabia* was thought completely lost.

As the years slipped by, unscrupulous businessmen later devised a scheme to sell some of the whiskey resting in kegs from the lost ship. Skillfully, under the cover of darkness, the men buried several barrels of freshly brewed whiskey. Then, in the light of day, barrels were dug up before a group of spectators. No one questioned the authenticity of the find. Its true worth was only considered as bar value and a ruse. The scoundrels later made claim they found the lost brew and resold the barrels at a premium price to unsuspecting saloon keepers. They in turn charged enormous prices to their unsuspecting clientele. The men later revealed, they found humor seeing bar patrons sipping the poor representation of whiskey claiming they were drinking the best Kentucky bourbon available. In truth, they never knew the difference.

Twenty years passed since the ill-fated sinking. A man by the name of Robert Treadway decided to recover the riches of the *Arabia*. For years after the ship went down, Treadway made detailed notes and maps showing the new path taken by the river. He was sure he knew exactly where the ship now rested. Then, in 1877, he convinced Frank Tobner, a wealthy merchant to finance the recovery of the precious Kentucky bourbon whiskey kegs. Treadway was certain the first barrels uncovered were fake and the original kegs still rested within the hull of the ship.

To his dismay, after four months of digging they reached the upper deck of the wreck only to find a wood crate containing several pieces of mud-soaked clothing, a far cry from the valuable whiskey kegs. Funds ran out and the search was disbanded.

Several more years passed. Treasure hunters who salivated over the ship's riches made added attempts. Each failed, showing nothing for their efforts.

In 1895, another group was formed. This time, a large steel caisson was devised, shoring the muddy walls to avoid cave-in. Digging continued, then a tube, was lowered onto the deck with hope of finding the valuable booty. Finally, breaking through the heavy wood planks, the hunters were elated thinking they found their treasure at last! Digging deeper into the bowels of the ship, the men only uncovered boards of precut lumber, more clothing, several cases of shoes, rotting leather goods and preserved jars of food. But, still no whiskey!

Disappointment prevailed. The spring thaw fell upon the salvage team. The nearby Missouri River was raising again adding difficulty to the recovery effort. It was then they decided to stop and return later in the year when the river subsided. The woebegone treasure hunters lost interest and never returned. The *Arabia* remained in her grave of silt still untouched.

It was nearly 100 years later when interest in locating the lost riverboat resurfaced. In July 1987, treasure enthusiast David Hawley, wielding a state-of-the-art magnetometer, relocated the carcass of the *Arabia* in a farmer's cornfield. The ship's remains were thought to be buried in 45 feet of river silt. Confident of his find, he and a crew of earthmoving equipment operators began to excavate the huge undertaking of retrieving the lost ship.

Making a massive effort utilizing bulldozers and large hydraulic shovels, the first outer ship components began to surface within the tailings. Her giant paddle wheel appeared, then, after added digging, the engine and its boilers. Finally, wood deck plates were removed, exposing hundreds of items to be used to service the neighboring river settlements.

With permission granted by the landowner, the Sortor family, excavation continued for several more months. Pieces of treasure were removed on a continuing basis. To the amazement of all, a majority of the items removed were still in excellent condition considering they had been buried in silt for almost a century and a half. Each piece gave a historical picture of what life must have been on the frontier.

It took another three years of digging to remove all the parts of the ship. Hundreds of pieces required identification, cleaning and preservation. It was decided to assemble all the components associated with the find and place them in a museum for display. The so-called "true-value" whiskey kegs were never found, only bottles of champagne retaining their bubbles under corks.

The *Arabia* Steamboat Museum can be found in Kansas City, Missouri, with hundreds of the artifacts found within her hull. It is a true picture of a time that passed and will be remembered.

Sea Search